A Practical Guide to Particle Counting for Drinking Water Treatment

A Practical Guide to

Particle Counting
for
Drinking Water
Treatment

Mike Broadwell

CRC Press
Taylor & Francis Group
Boca Raton London New York

CRC Press is an imprint of the
Taylor & Francis Group, an **informa** business

First published 2001 by Lewis Publishers

Published 2022 by CRC Press
Taylor & Francis Group
6000 Broken Sound Parkway NW, Suite 300
Boca Raton, FL 33487-2742

© 2001 by Taylor & Francis Group, LLC
CRC Press is an imprint of Taylor & Francis Group, an Informa business

No claim to original U.S. Government works

ISBN-13: 978-1-56670-306-2 (hbk)

**Visit the Taylor & Francis Web site at
http://www.taylorandfrancis.com**

**and the CRC Press Web site at
http://www.crcpress.com**

Library of Congress Card Number 00-032265

Library of Congress Cataloging-in-Publication Data
Broadwell, Mike
A practical guide to particle counting for drinking water treatment/Mike Broadwell.
p. cm.
Includes index.
ISBN 1-56670-306-9 (alk. paper)
1. Particle counting (Water treatment plants)
2. Drinking water — Purification. I. Title.
TD368 .B76 2000
628.1′62—dc21   00-032265
CIP

Preface

Particle counting is one of the most exciting and important technologies in drinking water treatment. Its benefits far outweigh the problems encountered when any relatively new technology is introduced into a new area of application.

It is my intent to provide in this book a comprehensive yet practical guide to understanding the technology of particle counting and its application to drinking water treatment — a book that will be useful to the plant operator as well as the consulting engineer who has to specify the equipment and incorporate it into the overall plant design.

The book consists of three parts. The first provides a broad overview of particle counting, including the basic principles of operation, applications in the treatment process, and the fundamentals of installation, operation, maintenance, data collection, and system integration.

Part II covers equipment specifications in detail. It provides the information necessary to make intelligent choices when selecting equipment for a given application.

Part III presents equipment currently available on the market, assessed in terms of the material covered in Parts I and II. It provides comparisons based on the technical specifications covered in Part II.

The necessary information is provided within these pages for making an informed, intelligent choice when selecting a particle counting system, and a guide for using the technology to the greatest benefit.

Mike Broadwell
Atlanta, Georgia

About the Author

Mike Broadwell has over 12 years experience in water treatment instrumentation and control, specializing in particle counting. He worked for two of the leading particle counter manufacturers as field applications engineer and product manager for drinking water treatment, during which time he worked with treatment plants and consulting engineers in all but a half dozen states in the U.S.

In addition to extensive field experience, he has directed the development of a line of particle counting equipment from inception, including some of the circuit design and engineering.

Mr. Broadwell provides independent consultation for particle counting technology and its application to drinking water treatment as well as marketing consultation for process instrumentation. He holds a degree in electrical engineering from the Georgia Institute of Technology.

Mr. Broadwell maintains a presence on the internet at *www.ParticleCount.com.*

Table of Contents

Part I

Part II
Understanding the Technology

Chapter 11 Computerized Data Collection

Chapter 12 Putting It All Together

Part III
Assessing the Equipment

Acknowledgments

This book represents the accumulation of knowledge and experience gained from many hours spent in treatment plants across the country, working with many interesting and knowledgeable folks. It would be impossible to even begin to list them all here. As for those more directly involved in the production of this book, I would like to thank Thomas Ginn, Jr., of the Cobb County–Marietta Water Authority for providing application data, and Bill Sandidge of Instrumentation and Design, Inc., for assistance in clarifying some of the issues with software and SCADA integration.

Most of the manufacturers have been generous in providing materials and information. Thanks to Dr. Holger Sommer of ART Instruments, Bob Bryant of Chemtrac Systems, Greg McIntosh of the Hach Company, and last but certainly not least, John Hunt of Pacific Scientific, with whom I first worked with particle counters, and who has been a steady source of encouragement throughout my career in the industry.

I would also like to thank my parents, John and Barbara Broadwell, for the foundation they provided to allow me this opportunity in the first place, and to my editors at CRC Press/Lewis Publishers, for their patience and perseverance.

Part I provides a broad overview of particle counting, including the basic principles of operation, application in the treatment process, and the fundamentals of installation, operation, maintenance, data collection, and system integration. It is intended to provide a foundation for understanding the more-detailed specifics of the technology, which are covered in Parts II and III.

CHAPTER 1

Particle Counting Basics

To apply particle counters properly, it is important to understand how they work. This chapter is intended to provide a simple overview. Part II covers the design and operation of the complete particle counting system in greater detail.

A. WHAT IS A PARTICLE COUNTER?

A wide variety of instruments are available for detecting and measuring particulate matter in liquids and gases. Similar terminology is used for instruments that vary greatly in operation. Only a few of the many types available have application for drinking water treatment. For our purposes, we are concerned with instruments that are used to measure microscopic particles in water.

1. Particle Counters vs. Particle Sizers

The two main types of instruments used for particle detection in water are particle counters and particle sizers. Confusion arises because particle *counters* also size particles and particle *sizers* count particles. Some people refer to counters as sizers and vice versa.

The difference lies in the particle concentrations each is designed to measure. A particle counter is designed to count particles individually, and therefore is designed for very low concentrations of particles. A particle sizer is used to measure the particle size distribution of slurries or other liquids containing a large concentration of particles. Particle sizers do not count particles individually. They calculate the counts indirectly.

For drinking water treatment, particle counters are used almost exclusively. Particle sizers have been used in some research applications for drinking water, usually to study higher-concentration raw waters. Sizers are several times more expensive and complex than the particle counters used for drinking water treatment, and are of little practical importance for everyday plant operation.

2. Types of Particle Counters

A particle counter is an instrument that combines a particle detection device, or sensor, with an electronic counting device. The sensor detects the particle and converts information about that particle into an electronic signal, which is then fed into the counting circuitry. Particle counters are chiefly distinguished by the type of particle sensor employed. Once the particle information is converted to an electronic signal, any of a number of types of counting electronics can be used to process that information. The types of counting circuitry available are covered in detail in Part II of the book.

There are two primary types of particle sensors that are used for water treatment. The most common uses light as the basis of measurement. The other uses electrical conductivity.

a. Light-Based Particle Counters

Virtually all the particle sensors encountered in drinking water applications are light-based instruments. This is primarily due to the fact that light-based instruments are the simplest and thus the least expensive instruments available.

b. Electrical Conductivity Particle Counters

A second type of technology is encountered in research applications, and perhaps a few treatment plants. This technology uses electrical conductivity to detect and size particles. These instruments are more expensive (and complex) by an order of magnitude over the typical light-based instruments.

B. PRINCIPLES OF OPERATION

1. Light-Based Instruments

Among the light-based instruments, two types predominate. Both use small laser light sources, commonly known as laser diodes. This laser light source is used to illuminate individual particles that pass through it. The difference lies in how the interaction between the light and the particle is measured.

Whenever light strikes an object, any of three things can occur. (1) Light may be reflected by the object. (2) Light may be absorbed by the object. (3) Light may pass through the object. This is known as refraction. (See Figure 1.1.) How much of each of these occurs depends on the material makeup of the object.

a. Light-Scattering Sensor

Light-scattering particle sensors measure the light reflected (scattered) by each particle. A detector that converts light to electrical energy (a photodiode) is used to measure the amount of light scattered by the particle. The detector is placed at an

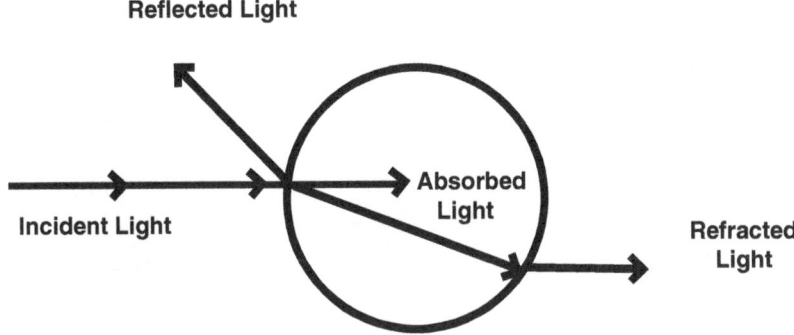

Figure 1.1 Light striking object.

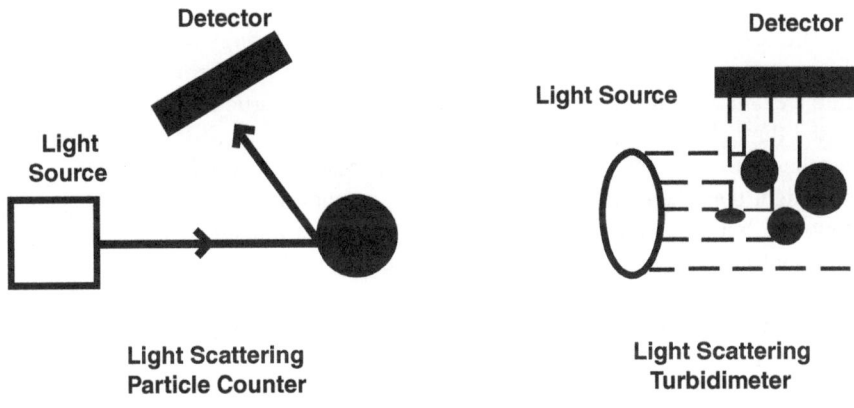

Figure 1.2 Light-scattering instruments.

angle of 20 to 40° from the light source, and will detect the light scattered at that angle by the particle. Particles are then measured or "sized" according to how much light they scatter. Larger particles will scatter more light than smaller ones. A similar form of light scattering is used in turbidimeter design (Figure 1.2).

b. Light-Blocking Sensor

Light blocking (also known as light extinction) measures the light absorbed or reflected away from the detector by the particle. In this arrangement, the light source is focused directly onto the detector, and the particle passes between them (Figure 1.3). It is obvious that the larger the particle, the more light it will block.

In drinking water treatment applications, light-blocking sensors are used almost exclusively. There are several reasons for this. The foremost is cost. Light-blocking sensors employ a much simpler optical design, which makes them easier to build and calibrate.

Light Source   **Detector**

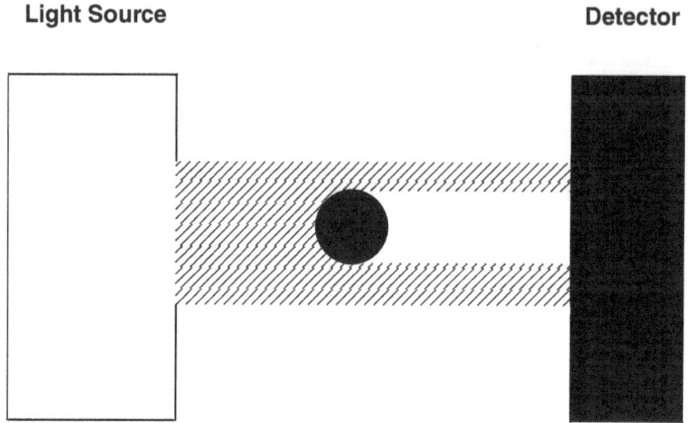

Figure 1.3   Light-blocking particle sensor.

The second reason for preferring light-blocking technology is that light-blocking sensors produce more consistent results with particles of different composition. A simple example will show why this is so. Suppose two particles of identical size are passed through each type of sensor. One of the particles is made of stainless steel, and the other of carbon. It is obvious that the stainless particle will reflect much more light than the carbon particle, and will appear larger to the scattering sensor. However, both particles will block almost the same amount of light. This is an important factor, as particles made of many different materials are found in water.

Both types of sensors will not properly size particles that refract light. As noted above, refracted light passes through the particle. Most organic particles have a low index of refraction, which means that they are transparent. This includes *Cryptosporidium* and *Giardia*, which are discussed in Chapter 2. Light-blocking sensors are less susceptible to this problem, as some light is refracted at an angle, and some absorbed. Less light is reflected back at the proper angle for a light-scattering detector.

Light-scattering sensors do have an advantage in sensitivity. Sensitivity refers to the smallest size particle that can be measured accurately. This is primarily due to the fact that scattering sensors detect light, whereas blocking sensors detect the absence of light. When no particles are present, the scattering detector is completely dark, while the blocking detector is fully illuminated. Just as it is easier to see a speck of light in a dark room than it is to discern that a speck has been extinguished in a brightly lit room, the scattering sensor can detect a smaller particle. However, light-blocking sensors can detect particles down to the practical limits necessary for water treatment.

Apart from these differences, light-scattering and light-blocking sensors are functionally equivalent. For the sake of simplicity, the rest of the book will focus on light-blocking sensors. Light-scattering sensors have been discussed because some will still be found in water treatment applications, and the phenomena associated with light-scattering discussed above are encountered to some degree with turbidimeters. It is also possible that light-scattering particle counters will again be used in drinking water treatment, if some new applications are developed that require the unique features of this type of technology. One potential application is the combination of light-blocking and light-scattering used to measure the same particles. This could be used to distinguish particles by material type.

c. The Rest of the System

Figure 1.4 contains a diagram of the complete particle sensor. A brief description is provided here to give the reader an introduction. A more exhaustive presentation of this subject can be found in Part II of the book.

To count and size individual particles, the particle counter must look at a very small volume of water. This is accomplished by using a laser light source to illuminate a small cross section of the sample stream. The sample stream is passed through an orifice approximately 1×1 mm. The height of the laser beam is around 1 mm. Thus, the active viewing area is 1 mm² or less.

The laser beam is passed through transparent windows on either side of the flow cell orifice. The beam illuminates the detector on the opposite side of the flow cell. This detector is used to convert light energy into electrical energy, which can be measured using electronic circuitry.

The sample stream is flowing at a fixed rate, usually around 100 ml/minute. As each particle passes through the laser beam, it reduces the amount of light striking the detector. The resulting output from the detector circuit is shown in Figure 1.5. Each particle produces a pulse output which is proportional in amplitude (height) to the size of the particle.

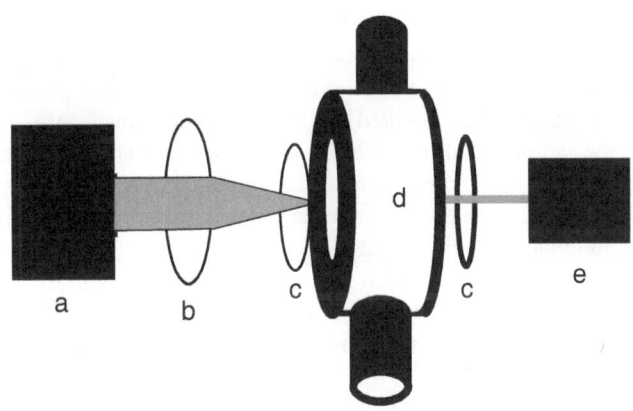

a. Laser Diode   d. Flow Cell

b. Focusing Lenses   c. Flow Cell Windows   e. Detector Circuit

Figure 1.4   Light-blocking sensor diagram.

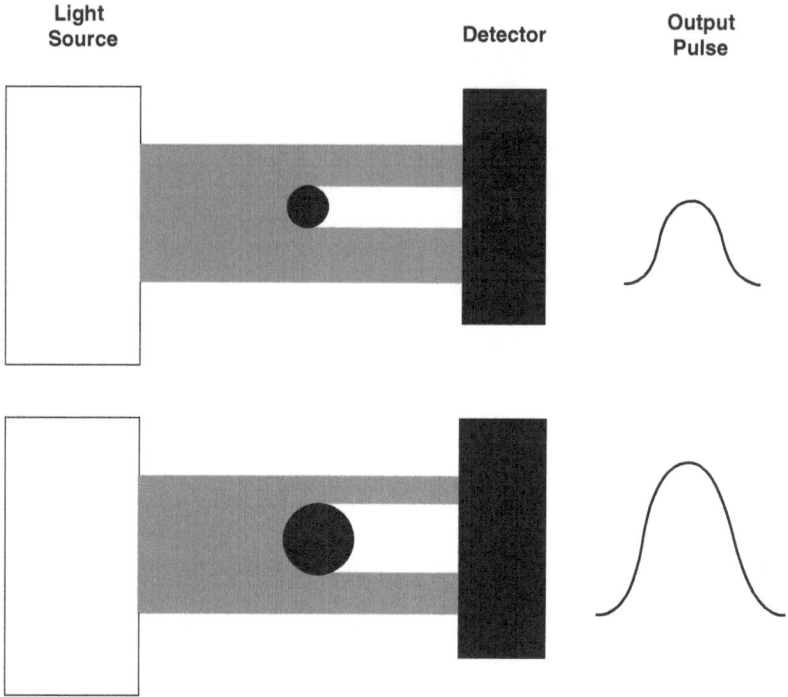

Figure 1.5 Detector output signal for differing particle sizes.

The output of the sensor is a stream of pulses, each corresponding to a single particle. The pulses are fed into the counting electronics, which sorts them according to size, and counts them for a fixed period of time. The flow rate of the sample is then factored in, resulting in a count per unit volume for each size range.

For example, if particles are counted for 15 seconds with the sample flow set at 100 ml/minute, the sample volume will be 25 ml (15 seconds is 1/4 minute, and 1/4 of 100 ml = 25 ml) The total number of particles counted over this time span must be divided by 25 to normalize the output to particles per milliliter. This calculation is usually done automatically by the particle counter.

The counts are divided into different size ranges, beginning with the lowest size particle that can be detected by the particle sensor (typically 2 μm). The number of size ranges varies between instruments, but four to six channels are typical for an online unit.

Several sources of error should be evident. Since the particle counter is looking at particles individually, there is a limit to the particle concentration of the sample. If more than one particle appears in the beam at the same time, the particle counter cannot distinguish them. This is referred to as coincidence error. The orientation of the particle as it passes through the beam will also affect how it is sized. Since this type of measurement is performed in two dimensions, the depth of the particle cannot

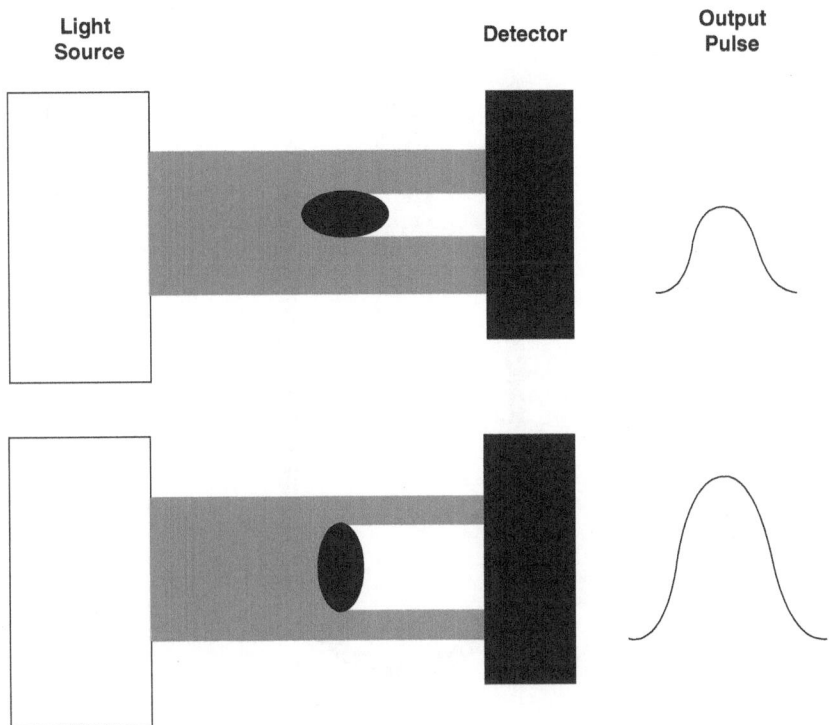

Figure 1.6 Variations in output pulse due to particle orientation.

be determined. An oblong particle can be sized differently depending on how it passes through the beam. See Figure 1.6.

Since particle counts are based on the volume of water sampled, the flow rate must be kept constant, or be constantly measured. Any error in the flow rate will translate into error in the output.

The laser light source provides a stable output, and is set up with a feedback circuit to maintain a constant intensity. Unlike conventional incandescent bulbs, there is no filament to degrade and shift over time. The laser diode will last for many years without degradation. This is the reason that particle counters can maintain calibration for a year or more under normal circumstances.

2. CONDUCTIVITY-BASED INSTRUMENTS

The conductivity-based particle counter is designed to measure the change in conductivity that results when a small, nonconducting particle displaces a small amount of electrolyte passing through an aperture between two electrodes. This type of measurement is known as the Coulter® method (Coulter is a registered trademark of the Coulter Corporation, Miami, FL); see Figure 1.7. The particles are suspended in an electrolytic solution that is drawn through a tiny aperture. The change in

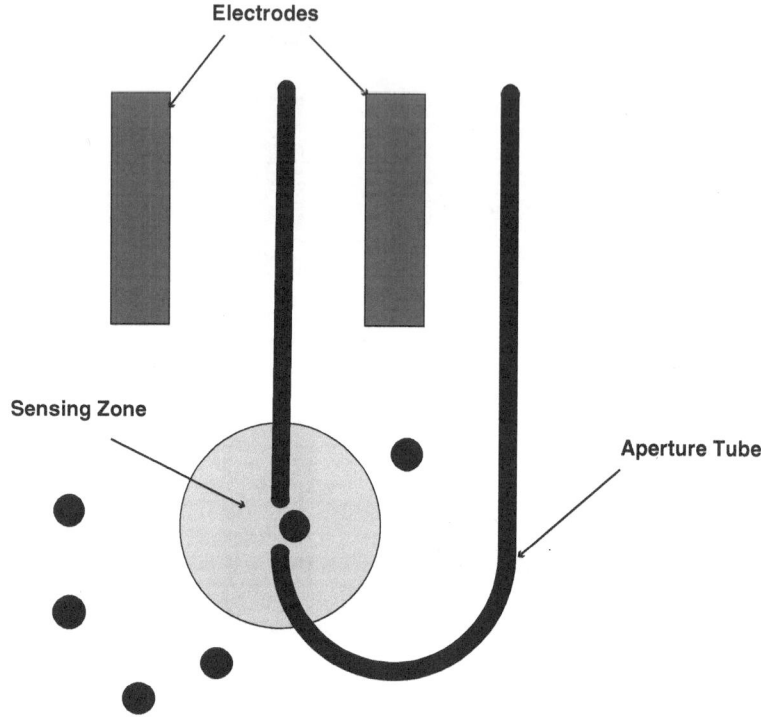

Figure 1.7   Coulter electrical sensing zone particle size measurement.

conductivity of the solution in the sensing zone is directly proportional to the volume of the particle.

This method of particle counting is primarily used for research applications, and is much more complex than the light-based methods presented in this book. It is noted because some drinking water research is done using Coulter counters.

A detailed description of this instrument is beyond the scope of this book. This instrument is much too complex for typical drinking water treatment applications. Since the particles must be introduced into an electrolytic solution, sample handling is relatively complicated. The Coulter counter will count and size particles well below 2 μm. It should be apparent that the Coulter counter is capable of greater accuracy because it measures the volume of the particle as opposed to the area measured by a light-based instrument.

C. FAMILIAR GROUND

Perhaps the best way to become familiar with using a particle counter in drinking water applications is to compare it with the instrument most closely related in both function and purpose. The turbidimeter is a familiar and useful device for measuring the particulate concentration in water. Most of us learned basic traffic laws by riding bicycles as children, and readily applied them to driving once the technical aspects of handling a car were mastered. Particle counters are used in the same places, and for the same reasons, as turbidimeters. Both a car and a bicycle are also used for the same purpose. It goes without saying that the car has some greatly superior features. But there are still places where a bicycle is more useful. The same holds true for particle counters and turbidimeters.

To stretch the analogy a bit farther, our goal in this book is to provide the small amount of technical training necessary to make using particle counters in drinking water treatment as familiar as driving a car.

1. Turbidimeters

Turbidimeters are used to measure the "turbidity" of a slowly changing volume of water. The word *turbid* comes from the Latin *turbidus* which means "confused." This is appropriate, since the whole concept of a Nephelometric Turbidity Unit (NTU) is somewhat confusing. About all one can say about a sample of water measured at 10 NTU is that it is 10 times as "turbid" as a water sample which measures 1 NTU. The concept began with someone named Jackson looking at candles through beakers of dirty water, not exactly what we would call "rocket science" today.

In spite of all this, the turbidimeter is a critical tool for evaluating drinking water quality, and the NTU has become the standard measure of finished water clarity. It is a big advance over the "eyeball" method in use since ancient times. The point was brought out for two reasons. The first is to show that we are comfortable with many things because they are familiar, not because they are simple. The second is to distinguish what is known as a *relative*, or *qualitative*, measure from an *absolute*, or *quantitative*, measure.

a. Relative Measurement

A relative measure is made by comparing (relating) one thing to another. This is necessary when the thing to be measured is a quality, such as "cloudiness" or "turbidity." An arbitrary standard is set, and becomes the basis of comparison for all other samples.

b. Absolute Measurement

An absolute measurement involves known quantities. This type of measurement can be made directly, without reference to other measurements of the same type. Particle counters are used to make absolute measurements. They are used to count particles and sort them according to size.

Both relative and absolute measurements are used widely in the water treatment process. One is not necessarily "better" than the other. The distinction is made to bring out the essential difference between the particle counter and turbidimeter. We will examine this distinction further.

2. Turbidimeter Operation

A turbidimeter measures the amount of light scattered at a 90° angle by the aggregate of particles in a certain volume of water. This volume of water may be constantly changing, as in an online turbidimeter, or it may be a fixed-volume grab sample; see Figure 1.8. The amount of light-striking the detector is proportional to the amount of particulate material in the sample volume. The light energy is converted to an electronic signal by the detector circuit. This signal is then measured and converted to the appropriate NTU value.

A turbidimeter can detect particles well below 1 μm in size. Any particles that scatter light can contribute to the overall turbidity. The sensitivity of the turbidimeter to these particles is not very high, so they must be present in significant concentrations. The amount of the sample volume is not critical for the turbidimeter to work properly. It is only important that the volume be constant. The turbidimeter is calibrated by placing a sample of known turbidity into the sample chamber, and adjusting the electronics to display that value.

There are several sources of error in this type of measurement. The light source must be kept constant, or the signal will not be accurate. Most turbidimeters use an incandescent light bulb as the source. These bulbs have a filament that will burn out over time, and will also shift slightly. As the filament ages, it will put out less light, and the shifting will affect the optical alignment of the instrument. This is the reason that turbidimeters must be calibrated often.

The problems inherent to light-scattering were mentioned above. Particles identical in size but made of different material scatter different amounts of light. Polymers can absorb scattered light, causing the turbidimeter to read a lower NTU value than it should. Carbon fines will scatter little or no light. Some of the light scattered by particles at the back of the sample volume can be blocked by particles closer to the detector. Color can also affect the turbidity reading.

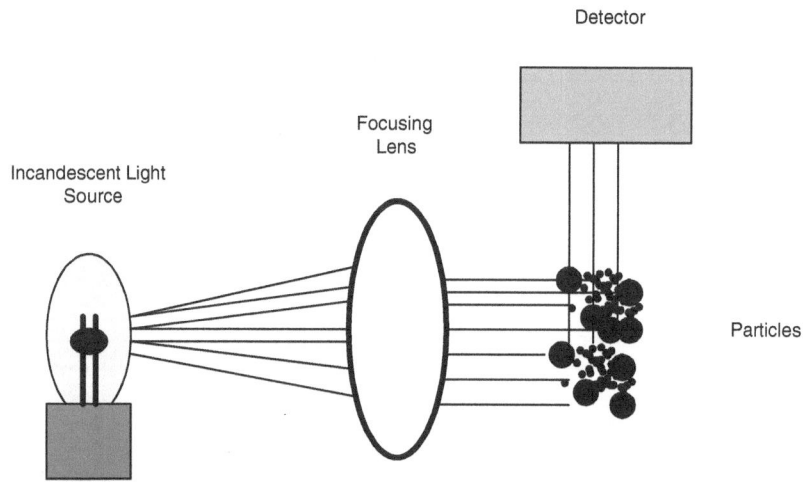

Figure 1.8   Turbidity sensor diagram.

1.07 NTU   **1.07 NTU**   **1.07 NTU**

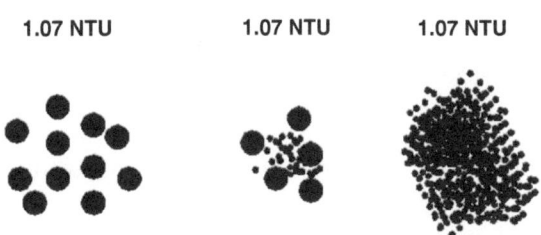

Figure 1.9 Various particle concentrations with identical turbidity readings.

All instruments have inherent errors and limitations. The important thing to remember is that the turbidimeter cannot distinguish particles *quantitatively*. There is no way to tell if a sample measuring 1.07 NTU is made up of a lot of small particles or a few large ones. Any number of particle concentrations can produce the same NTU reading. An example of this is shown in Figure 1.9.

3. Particle Counters and Turbidimeters

There are many similarities, and some important differences, between particle counters and turbidimeters. Most of these have been presented above. For the sake of clarity, we will summarize them again.

a. Similarities

Both instruments use a fixed light source to illuminate the particles suspended in water. The amount of light that interacts with the particles is measured with a detector circuit, which converts light energy to an electronic signal.

b. Differences

- Particle counters count individual particles according to their size. Turbidimeters cannot distinguish the amount or size of the particles.
- Particle counters provide a quantitative, or absolute, measure of the particles present. Turbidimeters provide a qualitative, or relative measurement.
- Particle counters are more sensitive to small changes in particle concentration than turbidimeters because they look at single particles.
- Particle counters use light-blocking technology, and turbidimeters light-scattering. Light blocking is less sensitive to the material makeup of the particle.
- Particle counters require a known, constantly flowing volume of sample. Turbidimeters need a constant volume only.
- Particle counters have a fixed sensitivity, and cannot count particles below a particular size (usually 2 µm). Turbidimeters can detect particles much smaller in size, provided they are present in sufficient concentrations.
- Turbidimeters provide a rough measure of the particulate concentration over a large range of particle sizes and concentrations, particle counters provide a more exacting measure over a limited range of sizes and concentrations.

- Particle counters use a laser diode light source, which will not degrade over time. Turbidimeters use an incandescent light source with a filament, which will degrade and affect the calibration. (Some turbidimeters are now being outfitted with laser diode or light-emitting diode (LED) light sources, but the majority are still incandescent.)

4. Particle Counters and Turbidimeters Are Complementary

For many of the reasons covered in the previous section, it should be apparent that particle counters and turbidimeters are complementary. Used together, they can provide a broad picture of the particulate content of the water throughout the treatment process. There are areas where both are not necessary, and, as the cost of particle counters drops closer to that of the turbidimeter, decisions on which instrument to use at a given point in the process must be based on performance considerations.

D. GRAB SAMPLE OR CONTINUOUS ONLINE?

Many process instruments used in drinking water treatment are available as continuous online units as well as laboratory instruments designed for grab samples. This is true of particle counters as well. The early appeal of grab-sample particle counters was based on the high cost of outfitting a plant with online units, as well as the uncertainty surrounding a relatively new technology. There was no point in spending a great deal of money on something that could be a passing fad. Since that time, costs have dropped considerably, and particle counting has been established as an integral technology for drinking water treatment. These developments have moved the emphasis away from cost, and toward operational proficiency and efficiency.

A grab-sample particle counter is basically an online unit with a built-in pump and flow regulator. A particle counter must have a known, constant flow rate to function properly, and this is accomplished by pulling the sample through the sensor with a pump. The major problems encountered are data management and sample handling, which are covered in later chapters.

CHAPTER 2

Applications for Drinking Water Treatment

This chapter provides an introduction to the application of particle counters in the drinking water treatment process. It is not intended as an exhaustive presentation, but rather as a starting point for looking more closely at the ways in which particle counters can provide valuable data for process optimization. Many different treatment processes and strategies are to be found in the drinking water industry, and source water quality varies greatly from region to region. It is hoped that readers will use this information as a catalyst for looking more thoughtfully and imaginatively at the particular application with which they are involved. It should also provide a framework from which to understand better some of the recommendations made elsewhere in the book.

A. WHY USE PARTICLE COUNTERS FOR DRINKING WATER TREATMENT?

A partial answer to this question has already been given in the preceding chapter. Particle counters are more sensitive to changes in particulate concentration than turbidimeters (in many cases), and thus offer additional information about process changes. The data presented below give some idea of the value of this sensitivity. Recent findings have indicated that treatment plants operated consistently with effluent turbidity levels below 0.1 NTU will experience few problems with waterborne pathogens such as *Cryptosporidium* and *Giardia*. The problem is that turbidimeter accuracy falls off greatly below the 0.1 NTU level. On the other hand, particle counters are tailor-made for these low concentration waters. They provide a much greater operating margin at these demanding treatment levels.

Particle counters detect particles in the size range of *Cryptosporidium* and *Giardia*, which is probably the major reason they have been so readily accepted into the drinking water industry. There has been a lot of misunderstanding about the way in which particle counters are used to combat these pathogens, which should be cleared up here.

At the most basic level, particle counters could not be a more natural fit for drinking water treatment. After all, water treatment boils down to two tasks. The first is to remove as much particulate matter as is practically possible. The second is to eliminate any harmful effects caused by the particles that cannot be removed. Particle counting is obviously directly related to the first of these tasks. As a further benefit, particle counters detect particles down to the size ranges below which removal becomes impractical for standard drinking water treatment. It is therefore no surprise that particle counting technology has been so quickly embraced in the industry, in spite of the technological shortcomings.

B. *CRYPTOSPORIDIUM* AND *GIARDIA*

A handful of major outbreaks of waterborne disease in recent years have been traced to the presence of *Cryptosporidium* or *Giardia* in the treated water supply. In most cases, this has been the result of process upset or operational error, which allowed these organisms to pass through the treatment plant unharmed. Few if any cases exist where a significant outbreak occurred while the treatment process was operating flawlessly. The problem comes with determining just how "flawless" is flawless, and with the awareness that it only takes one upset or breakdown or operator error to ruin a perfect track record. It is like the story of a troublesome employee who kept avoiding being fired by winning his union grievance hearings. His manager was nonplussed, stating that, "He's got to win every time. I've only got to win once."

Cryptosporidium and *Giardia* are parasites that live in the intestinal tracts of cattle and other mammals. They are spread into source waters by runoff from areas where these animals leave excrement. Untreated mountain streams are a source for these pathogens, as are lakes and reservoirs located near cattle farms or dairies. When ingested by humans, they can cause painful intestinal disorders sometimes referred to as "beaver fever," or "Montezuma's revenge." They can be fatal to infants or elderly people, as well as to anyone with a deficient immune system. The highly publicized outbreak in Milwaukee, Wisconsin in 1992 reportedly affected as many as 400,000 people.

Cryptosporidium is extremely nettlesome because it can survive fairly large doses of chlorine. To be effectively stopped, it must be filtered out of the treated water. Fortunately, it is large enough to be stopped by a properly operating conventional filter; see Figure 2.1.

C. PARTICLE COUNTERS AND *CRYPTOSPORIDIUM* AND *GIARDIA*

Particle counters used for drinking water treatment can detect particles down below the size of *Cryptosporidium* and *Giardia*. However, as noted in Chapter 1, organic particles are largely transparent, and thus will appear much smaller to the particle counter than they actually are. It is likely that *Cryptosporidium* will appear to be smaller than the 2 μm sensitivity limit of the particle counter. So one cannot rely on a particle counter to detect *Cryptosporidium* or *Giardia*.

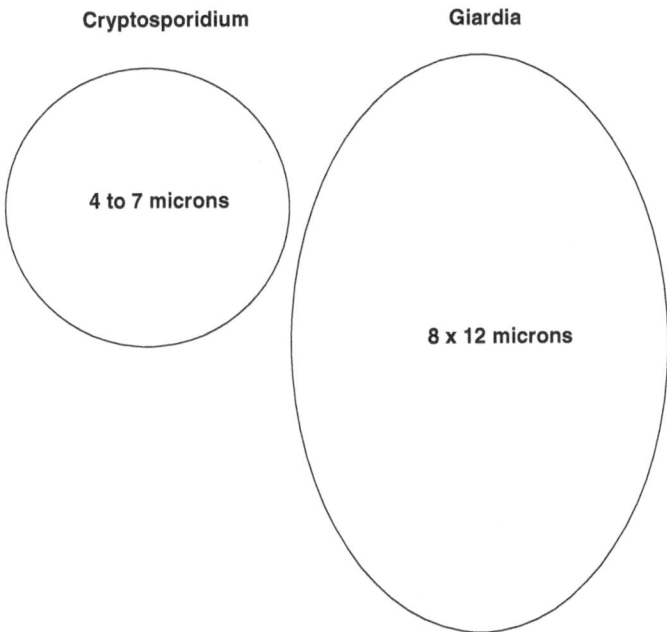

Figure 2.1   Typical sizes and shapes for *Cryptosporidium* and *Giardia*.

Furthermore, without directly referencing epidemiological studies, it is known that only one or two of these parasites is enough to cause illness in a certain percentage of the population. As more are ingested, a greater percentage of people will become infected. Let us assume that we have a situation where there are 100 active organisms per liter of water being produced. This should be well more than is needed to affect almost anyone (a dozen or more would be present in a single glass of water). Let us also assume that we have an ideal particle counter that can detect every one of them. Then, 100 Cryptosporidia/liter would work out to *one-tenth* of a particle per *milliliter*. If we had really clean filtered water to measure, we might see *less* than 10 particles/ml on average. Would an increase of 0.1 particle/ml make much of an impression on us? Of course not. It would not even be noticeable. So even if the particle counters could count the organisms accurately, it would not make any difference, except in extreme situations.

So why all the fuss about particle counters, if they cannot measure the very thing that they were brought in to combat? Why a whole book about particle counters?

D. SURROGATE MEASUREMENT

Particle counters are properly employed as a *surrogate* measurement tool. Surrogate means "to use in place of." Some may remember the controversy surrounding surrogate mothers a few years ago. These were women paid to carry children to term for women who were physically unable to do so. In our case, the less news-worthy surrogates for *Cryptosporidium* and *Giardia* are the other particles of the same size, which can be measured by the particle counter.

Particle counters are properly used to measure the removal efficiency of filters for particles which are the same size as *Cryptosporidium* and *Giardia*. It is assumed that if we can remove 99% of the particles we can detect with the particle counter, we are also removing 99% of those we cannot detect, i.e., *Cryptosporidium* and *Giardia*. To determine this removal efficiency, we must count the particles entering the filter and those exiting the filter. The relationship between these two values is usually referred to as the *log removal* or *percent removal* efficiency of the filter.

E. LOG REMOVAL

Removal efficiency is simply the ratio of particles exiting the filter to those entering the filter for a specified size range. This ratio may be expressed as a percentage, or logarithmically. The latter is known as log removal, the former as percent removal. Both represent the same value. Log removal is more widely used because it provides a much wider range for graphing values. For example, a log removal value of 2 is equal to a percent removal value of 99. Figure 2.2 gives an example of the reason it is easier to display values in log form. Log values are also used for chlorine contact time (CT) calculations. The two values can be added together to provide a combined removal and inactivation measurement.

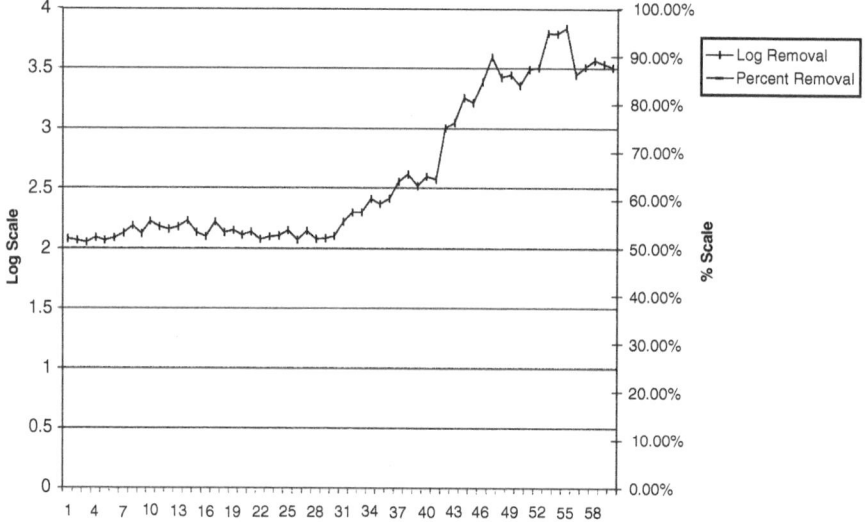

Figure 2.2   Log vs. percent removal.

1particle in effluent
100 particles in influent

1/100= .01

Log$_{10}$ (.01)= -2 = 2 log removal

(100-1)/100 = 0.99 = 99% removal

                     4 log = 99.99%
                     3 log = 99.90%
                     2 log = 99.00%
                     1 log = 90.00%

Figure 2.3   Log removal calculation.

Log removals are calculated by taking the log$_{10}$ (log base 10) of the number of effluent counts divided by the number of influent counts for a given size range. For example, one effluent count divided by 100 influent counts would equal 0.01. The log$_{10}$ of 0.01 equals –2. The minus sign is ignored (it is implied in the term *removal*), and we have a 2 log removal. It is easy enough to see that 1 out of 100 also equals 99%. The log$_{10}$ increments 1 unit for every order of magnitude. See Figure 2.3.

Much debate has centered around the use of log removal efficiency as a measure of water quality. The use of log removal as a regulatory guideline is questionable, for the fact that it is difficult to produce a good log removal value on low-count source waters, while filtering sewage through a wet rag might produce a 3 or 4 log removal. From an application standpoint, log removal is useful because it gives us a baseline for properly comparing filter performance. It is impossible to judge filter performance adequately over time without knowing the particle input as well as the output.

F. IMPROVING FILTER PERFORMANCE

As touched on briefly above, particle counters are most directly suited to monitoring filter performance. Filters are designed to trap particles down to 2 μm or less, and the particle counter affords a simple way of measuring how well the task is being accomplished.

The most basic application is that of determining whether the filters are performing properly. Since particle counters are much more sensitive than turbidimeters, they can show significant differences in filter performance, which will not register on the turbidimeter. This allows for an "early diagnosis" of problems that could have serious consequences if left unchecked. Consider the following example.

A treatment plant on the West Coast had recently installed two new filters that were loading up much more quickly than the four previously existing ones. Questions about the construction of the new filters arose, since the effluent turbidity levels for each filter were all well within acceptable limits. A couple of online particle counters

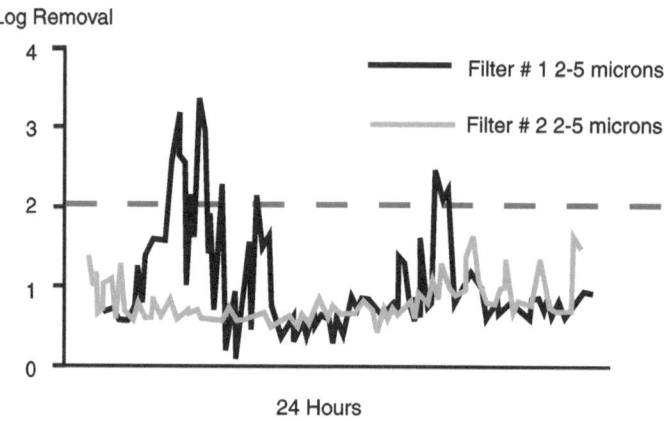

Figure 2.4 Old filter log removal. (Courtesy of Pacific Scientific Instruments, Grants Pass, OR.)

were brought in to allow a better look at the problem. Each of the six filters was monitored for about 24 hours.

The first two filters produced the particle count results displayed in Figure 2.4. These filters were part of the original plant design, and had never been rebuilt. These filters were performing quite poorly as can be seen from the extremely poor log removals in the smaller size ranges. A properly performing filter of this type should achieve at least a 2 log removal efficiency.

The second pair of filters was installed a decade or so after the plant was built. One was performing adequately; one was not. Figure 2.5 shows the results. While the first two filters were old enough to be worn out, Filter 4 was not. The media had been damaged, and it had been performing at an unacceptable level for who knows how long.

The data from the two newly installed filters are presented in Figure 2.6. It is obvious from the excellent performance indicated that they were not loading up too fast but merely working properly. Again, all of these filters were producing acceptable turbidity levels. The particle counters provided a truer picture of their performance, and, as a result, three of the four existing filters were rebuilt, and particle counters were installed on each filter.

Figure 2.7 shows data from a filter that had a small hole in an underdrain tile. The filter produced abnormally high counts when compared with the other filters. This was observed only on one half of the filter. While the particle counts did not directly point to the problem, they caused the operators to take a closer look at the filter, and the problem was discovered. Note that the counts on the faulty filter were still quite low, but were an order of magnitude higher than the other filter counts. This is a good example of why it is important to look for meaningful clues in the data, as opposed to targeting a specific number of counts.

Damaged filter media will often be indicated by carbon fines in the filter effluent. As mentioned in Chapter 1, turbidimeters will not detect these particles because

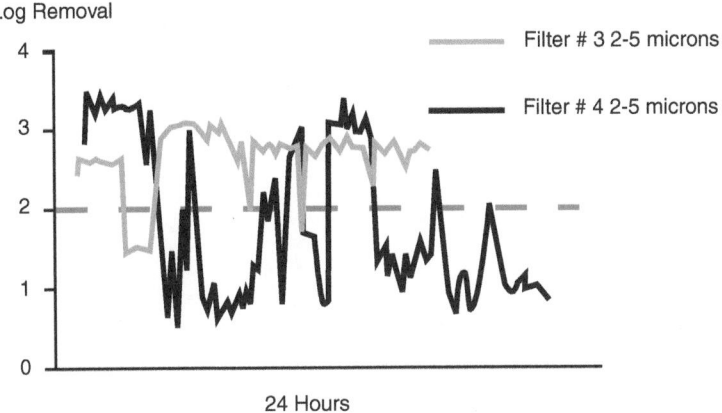

Figure 2.5 Middle-aged filter log removal. (Courtesy of Pacific Scientific Instruments, Grants Pass, OR.)

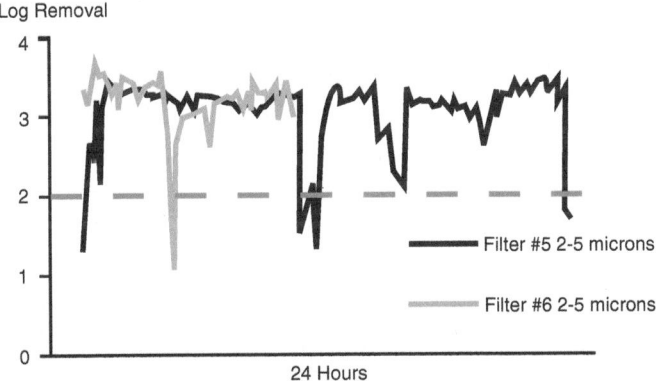

Figure 2.6 New filter log removal. (Courtesy of Pacific Scientific Instruments, Grants Pass, OR.)

they do not scatter light. They are easily detected by the light-blocking particle counter. While complete continuous monitoring of each filter is the most desirable approach, it is possible to diagnose potential problems with only one or two online particle counters.

1. Filter Run Time

Mechanically sound filters must still be operated properly to prevent particle breakthrough. Except for seasonal variation, most drinking water plants operate with consistent loading of the filters, so filter run times will remain constant. Particle counters will provide an excellent picture of the filter ripening process when the

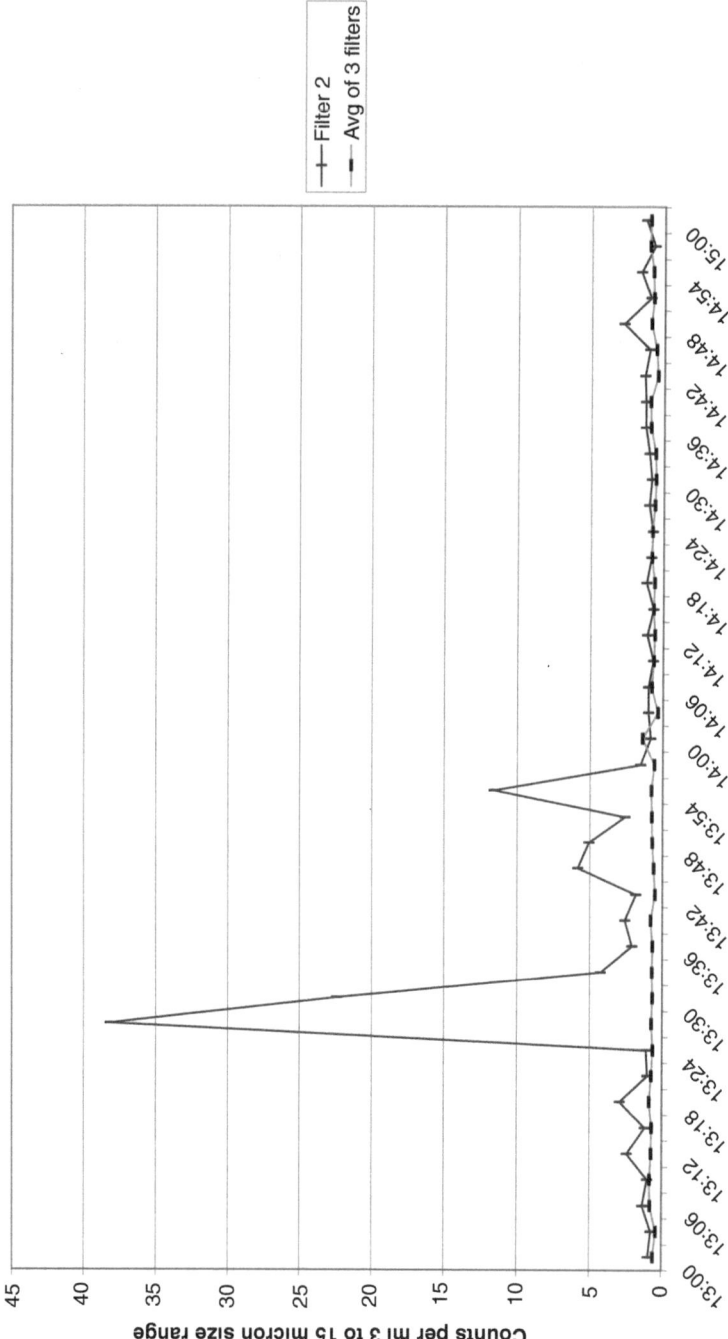

Figure 2.7   Damaged filter particle counts. *Note:* Damaged filter plotted vs. the average of the other three filters for clarity. (Data courtesy of the Cobb County/Marietta Water Authority.)

data are properly trended. They will quickly indicate increases in particles, providing early warning of potentially dangerous particle breakthrough. This high sensitivity to particle breakthrough is perhaps the most valuable attribute of the particle counter. Figure 2.8 provides a good illustration of this sensitivity. In this case, particle counts begin to move upward several hours before any change in turbidity is noticeable. Note also that the particle counts drop dramatically after backwash, whereas the turbidity drops more slowly. Filter-to-waste times can be adjusted for maximum efficiency.

Many factors affect filter performance. When a filter is removed from service for backwashing, the other filters will see an increase in flow. This will usually result in higher particle counts and shortened run times. Figure 2.9 provides an example of this.

A good technique for learning how to use particle counters is to learn to "tell time" from the data. Backwashing filters, turning pumps on or off, or any number of occurrences in the plant will produce spikes or other changes in the particle count data. The operator should be able to look at the particle count trend and trace the cause of any changes to various plant operations.

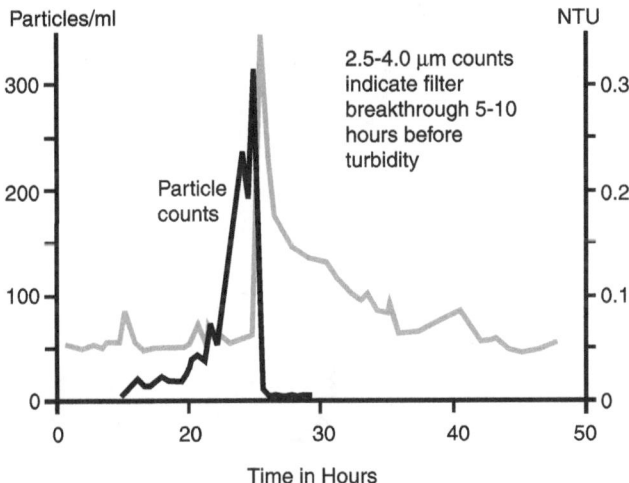

Figure 2.8 Particle counts anticipate filter breakthrough.

G. PROCESS OPTIMIZATION

The goal of proper drinking water treatment is consistent water quality at a cost-effective level. Like any real-world process, this involves trade-offs. Chemical additives are necessary, but excessive amounts can produce harmful by-products, and increase costs. Improvement in one phase of the process may cause problems in another. For example, polymers may improve flocculation but load the filters too quickly. Particle counters are not a simple solution to the many problems encountered in process optimization, but can add a helpful piece to the puzzle. This section will

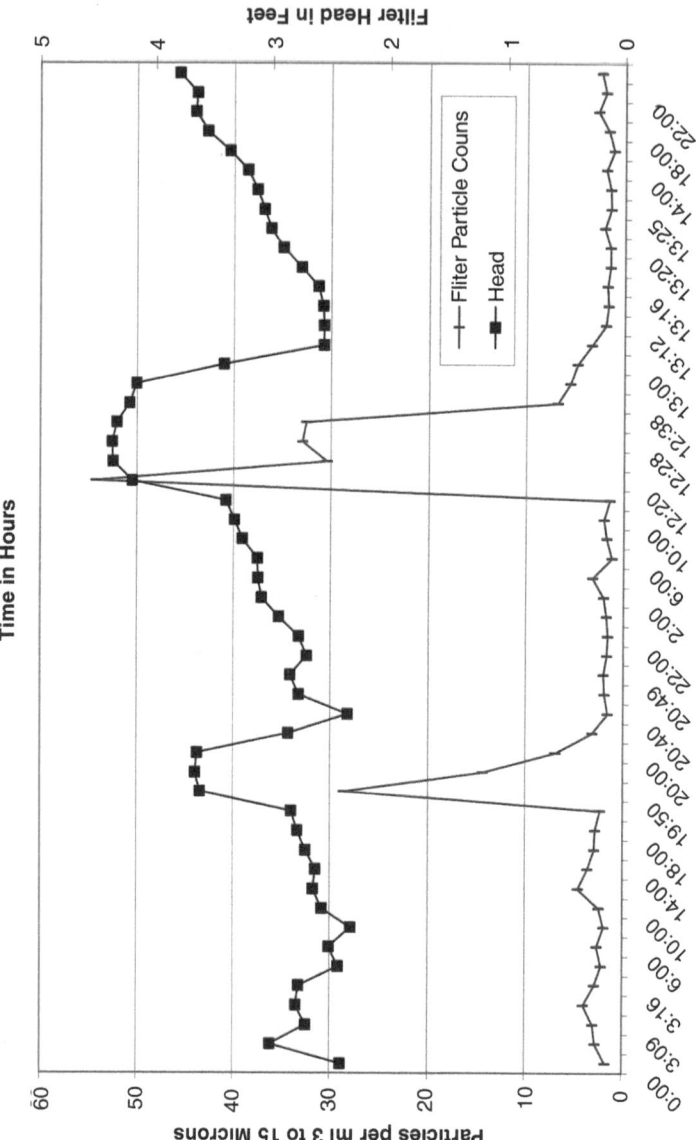

Figure 2.9   Particle counts vs. filter head. Increased loading as other filters are removed from service causes higher particle counts. (Data courtesy of Cobb County/Marietta Water Authority.)

look at a few of the areas where particle counters may be used to improve the treatment process.

1. Flocculation

Optimal filter performance is dependent upon proper flocculation. Since process conditions change seasonally, as well as for other reasons, it is important to monitor the effectiveness of the settling process. Particle counters can be used to measure the size distribution of the settled particles. This information can be used to determine the most effective chemicals and dosages for a given set of conditions. Chemical cost savings can be achieved, and unwanted by-products minimized. Longer filter runs will result from improved floc formation. Figure 2.10 shows particle count values as chemical feed is adjusted for improved efficiency. Note that the number of smaller particles is dramatically decreased. This is an indication of improved floc formation.

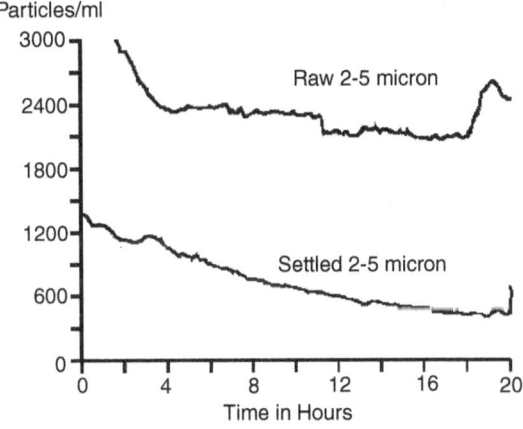

Figure 2.10 Flocculation efficiency. (Courtesy of Pacific Scientific Instruments, Grants Pass, OR.)

Table 2.1   Particle Count vs. Turbidity Pilot Plant Data

Particle Size Range	Alum Feed[a]			Polymer Feed[b]		
	4.0 ppm	5.5 ppm	7.0 ppm	Stabilized	Add Polymer (9 min)	Restabilize (30 min)
2–5 µm	128.86	41.06	8.35	69.20	6.25	35.44
5–40 µm	19.31	4.97	1.89	8.80	0.34	4.42
Turbidity (NTU)	0.14	0.055	0.045	0.1	0.08	0.06

[a] Alum feed results show that particle counts track with turbidity as dosage is increased.

[b] Polymer feed results show that particle counts show the opposite trend from the turbidimeter. (Courtesy of Pacific Scientific Instruments, Grants Pass, OR.)

Larger floc particles can break up when they pass through the particle counter, skewing the results. Particle counts cannot be sole criteria for setting chemical feed parameters. This is yet another area where a little imagination is required. The relative changes in particle counts, especially in size distribution, are important indicators of process change. Trending this data along with streaming current, loss-of-head, etc. will provide a good overall picture of process conditions.

Particle counters can also provide a "second opinion" to turbidimeters when analyzing various chemical additions. Polymers can fool turbidimeters into artificially low readings, which can lead to erroneous conclusions. The data in Table 2.1 give one example where the particle counter and turbidimeter tracked fairly well for alum feed, but gave contradictory results with a polymer.

2. High Rating Filters

One way to increase the output of a plant without building additional facilities is to high-rate the filters. Particle counters are almost mandatory for determining the acceptable rate at which a filter can be operated. Figure 2.11 gives an example of data collected during a rate test.

H. PROCESS APPLICATIONS

1. Conventional Treatment

The majority of particle counting applications will be found in conventional treatment plants. Conventional treatment incorporates the settling process mentioned in the previous section. In most cases, conventional treatment is employed where source water turbidities fluctuate over a fairly wide range. The settling process acts as a buffer to provide consistent loading for the filters.

In most cases, source or raw water particle concentrations will exceed the coincidence limits of the particle counter regularly. In such cases, it may not be practical to install a particle counter on the raw water. If turbidities exceed a couple of NTU only after heavy rains or on rare occasions, a particle counter may be useful. Dilution is practical with grab samplers, and online dilution systems are available, but use of them should be carefully considered.

The settled water should normally be well within the concentration limits of the particle counter, and should be monitored. It is often acceptable to measure the settled water at a single point, if the water is consistently applied to each filter. If separate settling basins are used to feed different groups of filters, then each basin should be monitored. The particle loading will usually vary between filters, as additional settling may occur before the water reaches the filters farther away from the basin. In such cases, it may be useful to take test samples from various locations to determine how much variation is encountered.

If at all possible, each filter effluent should have its own particle counter. Some plants will only install a particle counter on the combined finished water sample. While this is useful for measuring overall plant performance, it does not provide

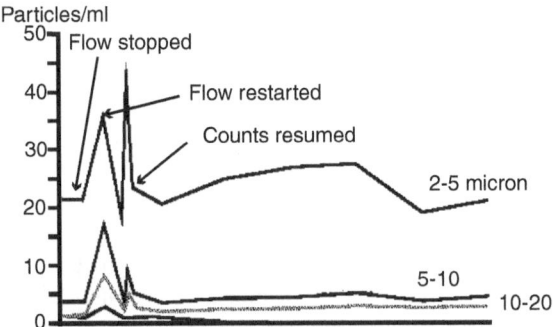

Figure 2.11 Filter high rate test. Flow started at 2.5 GPM/SF, then stopped and restarted at 2.5, then increased to 4. (Courtesy of Pacific Scientific Instruments, Grants Pass, OR.)

protection against individual filter breakdown. High counts from one filter may be diluted in the finished water to the point where problems may go unnoticed. At least one particle counter should be rotated among filters on a continuing basis, if all filters will not be outfitted. This will prevent a problem filter from operating for months or years before being detected.

2. Direct Filtration

Direct filtration is used for low-turbidity source waters. In these applications, the particle counter should be able to handle the concentration levels of the source water. Particle counters may be even more crucial in direct filtration applications, since there is no settling "buffer" to keep loading rates consistent, and there is less response time to deal with changing source conditions. Each filter should be monitored if at all possible, as any *Cryptosporidium* or *Giardia* occurring in the source waters could pass straight through a damaged or poorly operating filter. In all other respects, particle counters are operated identically as for conventional treatment.

3. Pilot Plants

Pilot plants run the gamut from conventional to experimental processes, so each case must be examined individually. Many treatment plants maintain a pilot plant to test process changes before applying them to the larger plant. In such cases, the pilot plant is designed to replicate the main plant. In these applications, particle counters should be applied in the same manner as in the larger plant. It may be desirable to sample other parts of the process not accessible in the main plant.

Often pilot plants are brought in to determine the best method to use in designing a new plant. Many premanufactured "packaged" plants are built in smaller communities. Often several manufacturers of these plants will be given opportunities to run pilot simulations to prove the effectiveness of their manner of treatment. In many states, particle counts are required in these applications. Most of the major manufacturers of packaged treatment plants employ particle counters as a standard part of their pilot plants. They are used not only to meet the state guidelines, but also to provide a more accurate measure of the performance of the system to help sell it to the customer.

Pilot plants are more difficult to operate consistently, for many reasons. Often, new or experimental methods are being employed, and the source conditions may be unknown. Usually the source water is pumped in from a river or lake, and the small size of the pilot plant makes maintaining the proper amount of throughput more complicated. Operational problems or mistakes may lead to the loss of a sale, or delay completion of the testing, adding additional costs to the project. It is important to characterize these new applications as accurately as possible, since the final installation will be a much larger and more expensive undertaking. These are some of the reasons particle counters are so important in pilot plant operations.

4. Membrane Plants

The last few years have seen an increase in the application of membrane filtration. Membranes are designed to remove particles above a certain size range, without the need for chemical additives. In these applications, the integrity of the membrane is all that stands between the source water and the finished product, with the exception of the chlorine added. Membranes are designed to be an absolute filter, i.e., to stop all of the particles above a given size. The source water is forced through tiny passages that trap the particles while letting the water pass through. If the integrity of the membrane is compromised, a large burst of particles will pass through. A small pinhole in the membrane material can become a source of thousands of particles, because of the pressure on the system.

Particle counters are used in membrane applications as a way to monitor membrane integrity. There is no ripening period as encountered in standard multimedia filters. The finished water is measured to look for rapid changes in particles, which would indicate a damaged membrane. Sizing is not important, as the membranes are designed to stop all particles above a minimum size. In most cases, particle counters are too expensive for permanent membrane applications. Membrane plants are normally for small applications, and several membranes are bundled together to produce the necessary throughput. Several points must be monitored by the particle counter to cover a membrane system adequately, but the cost of the particle counters is too large in relation to the cost of the membranes to make this a cost effective approach.

Membrane pilot plants often incorporate particle counters, for the reasons stated in the above section. In one instance, where an air backwash membrane system was being piloted, the bubbles produced by this process caused the turbidimeter to spike up for several minutes. A particle counter was brought in to ensure that particles were not passing through during this period. While the bubbles caused a brief rise in the particle counts, it only lasted for one or two samples, and the system was shown to be operating properly.

Most membrane systems incorporate internal pressure tests to determine the integrity of the membrane. Since these can be performed only every few hours, the possibility exists for significant breakthrough. Particle counters are an ideal solution to this problem, but until a trimmed-down low-cost approach can be developed, they will not be practical.

Reverse osmosis (RO) is a type of membrane application used for desalination and other problem source waters. RO processes usually involve several stages of prefiltration, because the final stage membranes are extremely expensive. It is important to remove as many particles as is practical before the final stage, to extend the life of these membranes. Particle counters can be used to monitor or troubleshoot problems in the prefiltration stages. In one application in a remote Arctic region, we encountered problems with the initial sand filter stage of an RO process. This filter was producing a large number of particles between 5 and 10 μm, which were passing straight through the 10 μm pre-filter and shortening the operating life of the RO membrane dramatically. While these concentrations were too small to impact the turbidity readings, they were causing significant problems. In such applications, a single grab sampler or online particle counter (or combination unit) is an excellent tool for troubleshooting.

5. Packaged Treatment Plants

Packaged treatment plants, as touched on above, are prefabricated plants, which are more cost-effective than ground-up plants for smaller applications. A packaged plant may involve any number of treatment methods, from conventional, to direct, to membrane. Many incorporate special processes, such as upflow clarifiers or dissolved air flotation, which can enhance the settling process. These processes often require changes in the way removal efficiencies are measured with particle counters. Most of the manufacturers of these special plants utilize particle counters in pilot plants, and can offer advice on how best to incorporate them into the final installation.

One of the major problems encountered is the lack of head pressure for a filtered water sample. Since these plants are prefabricated, no pipe gallery is built belowground. Special consideration will need to be given for particle counter application in these cases, much of which is covered in Part II of the book.

I. GROUNDWATER

Groundwater sources may be tested for surface water intrusion by using a particle counter to monitor for increases in particulate concentration that occur during and after rainstorms. Since the particle counter is much more sensitive to small concentrations of particles than a turbidimeter, it makes a better choice for this application.

Groundwater found to be under the influence of surface water may require filtering. In such cases, particle counters will be used as described above for conventional treatment.

J. WASTEWATER APPLICATIONS

Although not within the scope of this book, it might be of interest to look briefly at some of the potential applications for particle counters in wastewater treatment. Standard wastewater is too high in concentration for particle counters, but special-application areas hold promise. As the price of particle counters continues to decrease, interest should increase.

1. Tertiary Treatment

Standard wastewater treatment does not involve filtration. Organic waste is broken down with bacteria, and the effluent is chlorinated and discharged. In areas where this two-step process is not sufficient, a filtering stage is added. Filtration then becomes the third or tertiary stage of treatment. This filtration step is similar to that of conventional water treatment, and particle counters are used in the same manner. The particle loading in wastewater filtration is less consistent, resulting in filter runs of varying length. The particle counter is used to predict filter breakthrough as well as to spot problem filters.

2. Reuse

Tertiary treatment is employed in ecologically sensitive areas, as well as in drier areas where the effluent is used to water golf courses or other public spaces. This application is known as reuse. Some of the same concerns over *Cryptosporidium* and *Giardia* apply with reuse water, as it comes into contact with humans and animals. In addition to optimizing the tertiary filters, the potential exists for using particle counters to monitor points in the distribution system.

3. Ultraviolet (UV) Disinfection

The concern over disinfection by-products and their long-term effects has led to alternative means of disinfection. One of these is ultraviolet (UV) radiation. UV is an effective way to kill harmful pathogens and bacteria without chemicals. To ensure effective disinfection, sufficient doses of UV energy must be applied. The amount of UV required is proportional to the size and mass of the particles in the effluent stream. However, since UV generation requires a substantial amount of electrical power, to be cost-effective, the output levels must be continually adjusted to the particulate content.

Particle counters can be used to monitor the particulate concentration and provide a control signal used to raise or lower the UV dosage for maximum efficiency. This is especially of value when larger particles pass through the system. One concern is that living *Cryptosporidium* or *Giardia* particles will be clumped together with inorganic particles, which will shield them from the UV radiation. The UV dosage must be raised to a higher level to ensure disinfection of these larger particles.

CHAPTER **3**

Installation, Operation, and Maintenance

As with any instrument, proper operation will be achieved only if the particle counters are properly installed and maintained. This chapter covers the basics of particle counter installation, operation, and maintenance. The material presented is not specific to any particular make or model, but is intended as a general guideline. Model-specific information is covered in Part III of the book.

This chapter is primarily concerned with continuous, online particle counters. While much of it is relevant to grab-sample units as well, special consideration of grab-sample particle counters is given in Chapter 5 of Part I.

A. CHOOSING PROPER SAMPLE LOCATIONS

The most critical concern when installing particle counters is the proper selection of sample taps. The high sensitivity of the particle counter to microscopic particles makes it much more susceptible to error due to sample contamination than a turbidimeter. Care must be taken to minimize sample errors if accurate data are to be collected with the particle counter. Fortunately, proper sample tap selection requires little technical expertise outside of familiarity with the treatment process and good old-fashioned common sense.

We say fortunate, because one cannot become an expert at particle counting without using particle counters for a while, and they cannot be used until they have sample flowing through them. This is not a trivial point. When particle counters are first installed, there is no baseline or simple check to ensure that they are "working right." No green light saying "OK" will appear. The only confidence that the operator can have that the units are working properly is that careful and thoughtful attention has been given to every detail of the installation process. This is especially true for the sample connection.

As stated in the beginning of the book, 90% of the knowledge required to operate particle counters in drinking water treatment plants is already understood by a competent operator. Let us briefly review the basics of good sample tap selection common to all process instruments, and then add in the 10% of additional information required for particle counters.

1. Representative Sample

The sample must be representative of the process. This point is obvious enough. Most instruments only sample a tiny fraction of the process stream, and if that small sample does not reflect the overall stream accurately, it is not only useless, but could result in errors that adversely affect the whole treatment control process. The most representative point is usually in the center of the process stream. Here the velocity is highest, providing the most up-to-date changes, and the sample is most evenly mixed.

Figure 3.1 shows four possible tap locations for a turbidimeter sample. The requirements for particle counter sample taps are the same as those for the standard turbidimeter. Note that the bottom and the top of the pipe make poor choices, because of air and sedimentation. The side of the pipe is a better choice, but if the sample tap does not extend well into the pipe, it will not reflect the process accurately. Particles tend to cling to the walls of the pipe, and will release periodically, artificially increasing the count totals.

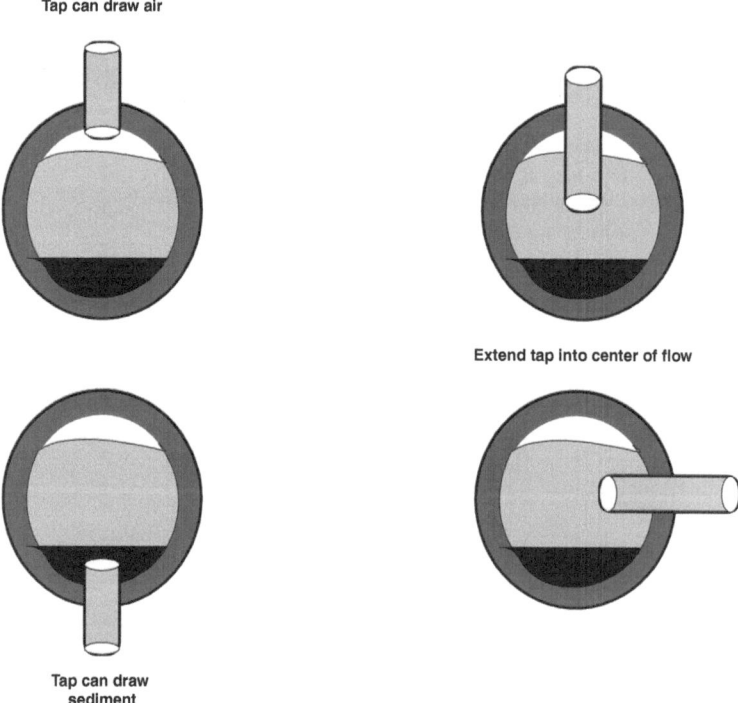

Figure 3.1 Sample tap location.

The same guidelines would hold true for settling basins and reservoirs. The suspended particles are the ones that will pass on to the filters, not the floating floc particles on the surface or the larger ones that sink to the bottom.

The differential pressure transducer is often used as a sample point for settled water. It is important to avoid the "mudleg" of this device because of the excess particles that lodge there.

2. Short Sample Lines

Keep the sample lines short. This is standard practice for most process instruments. Short lines keep the sample representative, prevent particle drop out, and minimize temperature changes, which can result in bubbles coming out of solution. Particle counters are sensitive to what is known as "particle shedding," a periodic buildup and release of particles from the walls of the sample tubing. This can result in intermediate bursts of particles that do not accurately reflect the process. Obviously, the longer the sample lines, the more surface area available for this shedding to take place.

The sample lines should be no longer than 10 to 20 feet.

3. Sample Line Materials

Several materials are available for sample lines, the cheapest and most practical being synthetic flexible tubing. The most commonly used tubing is the transparent Tygon™ tubing, which is inexpensive and readily available. Tygon does collect particles along the walls and discolors readily when chlorine and other chemicals are present.

Teflon tubing does not collect particles as readily, but is a good bit more expensive. It is less flexible than Tygon. Black nylon tubing should be used in areas exposed to direct sunlight. Transparent tubing is susceptible to organic buildup when exposed to sunlight. The drawback to black tubing is that it is impossible to determine the amount of particle buildup inside.

4. Valves, Pumps, and Manifolds

In most cases, it is necessary to place a valve on the sample tap. This allows the tap to be shut off when the instrument is removed from service. In such cases, a ball valve should be used. Ball valves are less prone to particle shedding than other types because of their smooth, rounded surfaces.

Pumps should be avoided whenever possible because they not only shed particles, but break up the particles in the sample. This can skew the count and size distribution. If a pump is necessary, it should be used downstream of the particle counter. This allows for the particles to pass through the particle counter before being altered by the pump.

Some particle counter installations use the samples pumped up to a central laboratory. While these lab areas are convenient for many measurements, they are not desirable for particle counters. This approach should only be taken if no alternative exists, since this type of sampling arrangement violates virtually all the established guidelines for proper installation. Particle counters are designed to be mounted in the pipe gallery, close to the sample taps, and the convenience of having them all together in one place does not outweigh the downside. Some older plants leave no alternative, but a new design should never incorporate this approach.

Some users have investigated manifold systems, where several sample lines are switched through a single particle counter. This approach was impractical back when particle counters were a good deal more expensive than they are now, and as the prices drop for particle counters, it makes even less sense. As in the case of laboratory pumps, manifold systems violate every good practice for sample handling. To run several samples through one particle counter, the lines have to be run all over the plant, extra valves are necessary, and a whole host of complications can arise.

5. Temporary or Shared Sample Locations

Many cases arise where particle counters are to be used only temporarily in a plant, or moved to locations in the plant. These might involve equipment evaluations or short-term troubleshooting of a filter. In such cases, it may not be desirable to install permanent sample taps.

Many of the locations will already have sample taps for other instruments. It may be possible to split off a sample line for the particle counter from these taps. In such cases, a "Y" shaped fitting should be used instead of the more common "T" fitting. The sharp right angle in the "T" fitting can cause the larger particles to split off, skewing the particle distribution of the sample. Care must be taken not to alter the sample flowing to the existing instrument. Make sure that the makeup or volume of the sample is not changed in a manner that will affect it adversely.

It is not advisable to take the sample from the effluent of the existing instrument, as the particle concentration will likely be altered. It would be better to pass the sample through the particle counter first, since the particle counter will not chemically alter the sample. This is still not a good practice, as periodic cleaning of the particle counter will probably cause problems with the other instrument. Unless no alternative is available, split the sample instead of passing it through both instruments in series.

The Hach 1720C turbidimeter provides a good source for a temporary sample. It has a constant-head sample chamber, which provides easy access to the sample. Pass the particle counter tubing down into the reservoir and then siphon the sample to start the flow. It may be necessary to increase the flow to the turbidimeter to maintain the proper level in the reservoir.

If a temporary sample is needed from a settling basin or reservoir where no taps are accessible, the particle counter tubing can be dropped into the basin, and either siphoned out or pulled out with a pump located downstream from the particle counter. A small weight should be attached to the sample tubing to cause it to sink a few feet below the surface. It should be kept away from the bottom or sides of the basin, and below the surface to avoid pulling air or floating floc particles.

6. Practical Considerations

In most cases, less-than-ideal conditions exist for choosing tap locations and minimizing sample line lengths. For instance, the shortest line length may require that the particle counter be mounted behind a pipe where it is hard to access. If it is hard to access, it will not be cleaned and maintained properly, and will eventually be ignored or taken out of service. It is much better to mount the particle counter where it can be easily reached for maintenance, even if the sample line length is increased. Conversely, the best mounting location may require an excessively long sample line. Perhaps no electrical power is available at the best location, and a great deal of expense will be required to complete the installation.

No two treatment plants are alike, and the approach taken will vary with the circumstances. It is always possible to experiment, perhaps by mounting the particle counter on a sawhorse and moving it around to different sample locations to test the results. It may well be that a much more convenient location will not affect the performance significantly. If nothing else, such experiments will help operators gain valuable experience with the particle counters.

B. SAMPLE FLOW

Just as the particle counter is extremely sensitive to sample contamination, it also requires a stable and constant sample flow rate. The reason for this should be obvious: particle counter data are expressed in particles per milliliter. Particles are counted for a specific volume of water. Just as sample contamination will skew the number of particles counted and create erroneous data, changes in flow that are not accounted for will create errors due to counting particles over too large or too small a volume of sample.

Fortunately, flow and flow control are areas with which the water treatment operator is well acquainted. No knowledge of particle counters is required to understand all there is to know to set up and maintain a proper flow control system. Unfortunately, this is the area where most of the problems occur in particle counter operation. The small orifice and sample flows necessary for particle counting account for the added difficulty. However, these factors do not make the problems more difficult to understand. They just require a little more forethought and attention to detail.

In short, there is no excuse for particle counter flow problems. Complicated flow systems are seldom required for typical water plant applications. Following the steps outlined below should prevent most of the problems encountered without adding a lot of unnecessary expense.

1. Maintaining Constant Head

The most important aspect of keeping the sample flow constant is maintaining a constant-head pressure on the particle counter sample inlet. All particle counters have tiny flow cells — usually on the order of 1 mm by 1 mm or smaller. Since

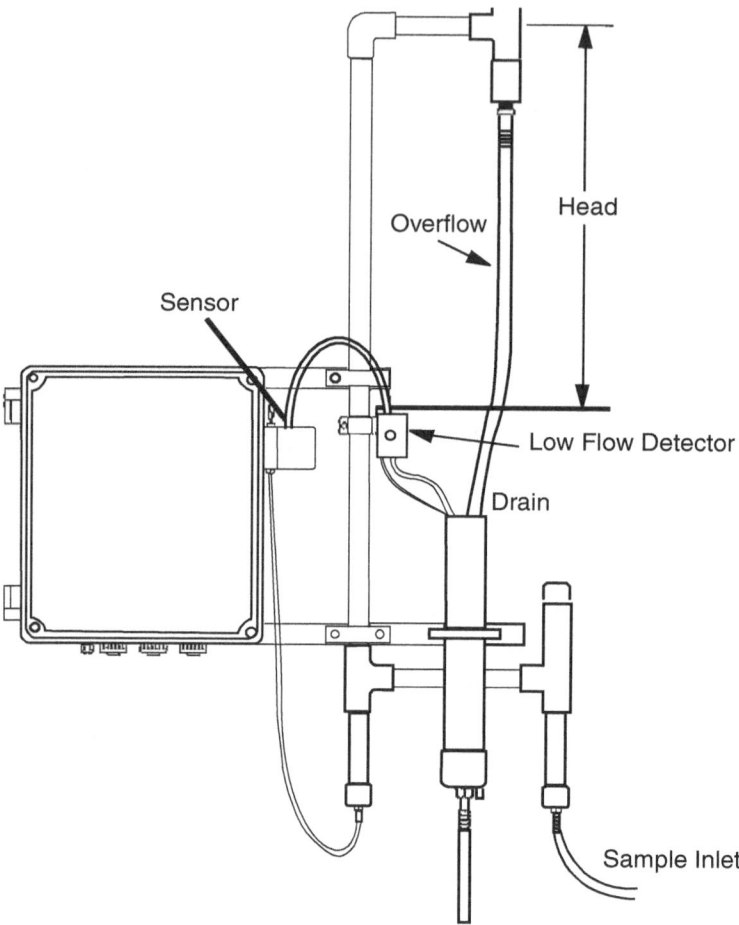

Figure 3.2   Constant-head overflow weir. (Courtesy of Chemtrac Systems, Inc., Norcross, GA.)

flow rate is directly proportional to pressure, the flow will increase or decrease along with the pressure. It is also obvious that the smaller the flow channel, the more flow will increase in proportion to the change in pressure. Thus, even small pressure changes will cause large changes in flow through the particle counter. The only practical way to prevent this is with a constant-head flow controller. These are inexpensive and are usually supplied with the particle counter. See Figure 3.2.

Constant head is maintained by use of an overflow weir. Flow is held constant by the constant-head pressure maintained by the weir. Pressure changes at the inlet are offset by proportional changes in the amount of overflow.

The long overflow tube is used to maintain enough head pressure to prevent bubbles from coming out of solution. The height of this tube is not directly proportional to the flow rate. Rather, the head height is measured from the overflow point to the sample outlet. The height of the flow cell relative to the overflow and outlet points is not critical. The head height is only dependent on the two points open to atmospheric pressure. Once the constant-head overflow weir is mounted, the sample outlet is raised or lowered to the height that will produce the desired flow.

2. Mounting the Constant-Head Overflow Weir for Best Operation

To achieve the desired flow at each sample location, the constant-head overflow weir must be mounted at the proper height. Ideally, this height would also be one convenient to access for periodic maintenance. The first rule of thumb is that the constant-head overflow weir must be mounted so that some overflow exists at all times, with the particle counter connected and the outlet tube set to produce the desired flow rate. In most cases, the greatest care will have to be taken with the filter effluent mountings. This is because the filters will experience several feet of headloss during a typical filter run. The constant-head overflow weir must be set up so that it operates properly at the maximum headloss of the filters.

Some filter galleries have only a few feet of space to work with, and some have filter effluent taps only a couple of feet off the floor. Even at minimal filter headloss, there may not be enough head to operate the weir.

In these cases, the constant-head overflow weir may need to be shortened, or the flow through the particle counter reduced. Consult the particle counter owner's manual before changing the flow rate through the particle counter, to determine the acceptable limits. The relation of flow rate to performance is discussed in Parts II and III of this book.

Most of the other sampling locations, such as settling basins, clear-wells, and reservoirs, are kept at fairly constant levels. For these locations, set the constant-head overflow weir for enough overflow to allow for some variation. In most cases, a ball value can be used to regulate the amount of sample flowing into the weir. There is no need to waste an excessive amount of sample, and too high a flow into the weir will exceed the limits for which it can maintain constant head.

New plant designs should take into account the flow requirements for particle counters, and provide for sufficient head (and space) to allow the units to be mounted at a comfortable working level for maintenance.

3. Other Flow Devices

In most typical plant installations where attentive maintenance is practiced, the constant-head overflow weir should be sufficient for controlling the sample flow. There are some cases where flow-monitoring devices are required, whether to provide better safeguards or to compensate for poor maintenance practices. Several options exist, and will be discussed briefly.

a. Direct-Reading Rotometers

The flow through the constant-head overflow weir is usually measured with a graduated cylinder and stopwatch. Direct-reading rotometers can display the flow without requiring this step. They are useful for performing quick checks to make sure the flow has not changed. It is important to remember that low-cost rotometers are only accurate to 5 or 10% of full scale, which can be a significant amount. They can also clog up on settled and raw waters, causing a drop in flow.

Needle valve rotometers should not be used to regulate flow. The needle valves will clog up quickly, especially when polymers are in use. The constant-head overflow weir should be used to regulate the flow, and the rotometer to read it. It is still necessary to check the flow periodically with a graduated cylinder and stopwatch, as the rotometer can produce inaccurate readings. We have seen one case where the rotometer ball was sitting at exactly 100 ml/min, while almost no sample was flowing out of the particle counter. The flow looked correct on the meter, and no one had bothered to check it.

b. Low-Flow Detector

A useful device for monitoring flow is a low-flow detector. This is usually attached to the constant-head overflow weir to monitor the flow out of the particle counter. It is set to sound an alarm if the flow drops below a certain point. Since most problems occur because of drops in flow due to clogs or excessive headloss (the constant-head overflow weir prevents flow from increasing) the low-flow detector will detect most flow problems.

c. Electronic Flowmeters

Many types of electronic flowmeters have been tried on particle counting systems. Few are practically feasible. The less-expensive types use a turbine wheel, which is susceptible to clogging. Most meters of this type are designed for particle-free liquids. In most cases, the sample must be filtered before passing through a turbine-type meter. This requires that a filter be placed between the particle counter and the flowmeter. This filter creates headloss, and must be replaced periodically. The nonintrusive-type meters tend to be more expensive, some costing more than half as much as the particle counter. These meters are usually designed to handle special chemicals, and are often made of materials designed to handle corrosive or high-purity liquids. This drives the cost up even farther. The very low (100 to 200 ml/min) sample flow rates used for particle counting are difficult to measure, and it has only been feasible for applications where the cost of the process can justify expensive instrumentation. Needless to say, drinking water is not one of the them.

Tritech Enterprises of Grants Pass, Oregon, has recently introduced an electronic flowmeter designed specifically for online particle counters used on drinking water sources. It is designed for flows ranging from 40 to 120 ml/min, and guarantees 1% accuracy. It is designed for use with the constant-head overflow weir, and is easy to install. This flowmeter is in effect an automated graduated cylinder and stopwatch, using a microprocessor-controlled timing circuit and a solenoid valve to fill and flush a constant-volume chamber. It mounts downstream of the particle counter, and does not create a pressure drop. The flow path is larger than the sensor flow path, so clogging is not a problem.

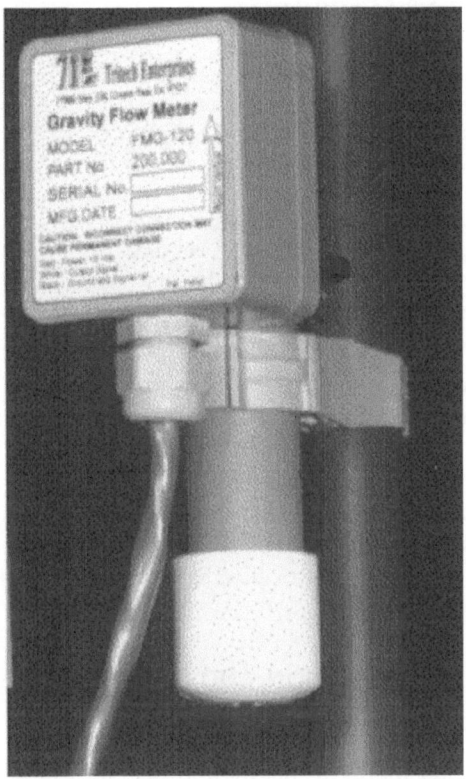

Figure 3.3 Tritech electronic flowmeter. (Courtesy of Tritech, Enterprises, Grants Pass, OR.)

This meter provides a 1 to 5 v DC output, and can be directly interfaced to most standard particle counters. It is compatible with any particle counter that operates within its flow range, and has the capability of reading an analog input signal. See Figure 3.3.

d. Determining the Best Approach

There is no easy fix for maintaining optimal sample flow outside of intelligent application of some very basic principles and vigilant maintenance. Adding expensive flowmeters because the maintenance staff cannot be counted on to monitor the particle counters properly is not the best solution. Most of the problems that will affect sample flow have to do with obstructions in the flow path of the particle counter, and this will still occur with the flowmeter installed. A flowmeter is yet another device that will have to be maintained and calibrated. The wrong flowmeter can create more maintenance problems than it solves.

Some particle counters come equipped with flowmeters or low-flow alarms as standard equipment. Others can be added as options. It is well worth the trouble to determine the added cost of these items before specifying a system.

We recommend the following approach to determining the proper type of flow-monitoring equipment for a particle counting system.

1. Use a constant-head overflow weir for each particle counter, regardless of the flow-metering equipment used. Particle counters are designed to operate over a narrow flow range, due to factors that are covered in Part II.
2. If a low-cost low-flow alarm is available, include it in the system.
3. Install the system without flowmeters, and determine how many, if any, problems with flow control are encountered. Have the flowmeters quoted separately and reserve the right to purchase them at a later time.
4. Place flowmeters only on the most troublesome units. There is no need to install them on every particle counter if only a few are causing problems.

For example, filter effluents should be easy to maintain because there are few particles to clog the particle counter, and most headloss problems can be solved with the constant-head overflow weir. Filter effluents constitute the bulk of the installations in most plants, so a big savings can be realized.

Once experience is gained with the particle counting system, it becomes easy to spot flow problems from the data. Even if flowmeters are added at a later time, the initial operating experience will provide a baseline for evaluating their usefulness.

C. OPERATION AND MAINTENANCE

Proper maintenance is essential to the operation of any instrument. Two factors will determine how much effort is put forth to keep up a given piece of equipment. The first is the relative importance of that equipment to the plant operation. The second is the amount of time and effort required to keep it in working condition. These factors are interrelated. Essential equipment will be maintained regardless of the effort required. Nonessential equipment will be kept up as time and resources allow.

Particle counters are often regarded as nonessential to plant operation. This is because the particle counter data are usually not reported to regulatory agencies, and most of the plant operations staff does not understand particle counters or how they are used.

The best way to keep particle counters maintained and operating properly is to provide the staff the training to understand the importance of the particle counting system to plant operation, and to hold them accountable for keeping it up. Most problems are flow and sample related, and are not complicated. Problems related to data collection and computer interfacing will have to be handled by someone with more specialized training. Data handling and computer maintenance will be discussed in the sections of the book related to that subject.

1. Maintenance Schedule

All good maintenance programs operate around a routine schedule. Regular checks of each particle counter should be performed, whether daily, weekly, or biweekly. How often checks should be made will depend on the conditions unique to each plant. Daily checks should be performed on a newly installed system. After a few weeks of familiarity with the system, the time between maintenance checks can be extended.

Routine maintenance checks are used to prevent problems that will occur due to buildup over time. For instance, a particle counter used to sample settled water may experience flow problems due to floc buildup every 3 or 4 days. To prevent this from becoming a problem, the particle counter should be flushed every day. If the buildup causes problems on a daily basis, it should be addressed on every shift. The particle counters sampling filter effluents may never experience flow problems due to continuous buildup. A routine weekly or biweekly check will be enough to keep them running properly. Focus the maintenance effort on the problem areas, and do not take up time unnecessarily on the rest.

The maintenance schedule should account for seasonal changes, and be adjusted accordingly. Permanganate is often added seasonally, and can cause coating problems. Any change in the process that can affect flocculation or filter loading should be taken into account. Raw water intakes may be shifted seasonally to avoid algae, changing the particulate content at the influent. Significant process changes should be treated the same as a new particle counter installation, with daily monitoring of the particle counters until experience warrants otherwise. Often the particle counters will be used to determine the effectiveness of process changes, and if they have quit working properly because of these changes, valuable information will be lost.

2. Unscheduled Maintenance Problems

The routine maintenance schedule should minimize problems due to buildup over time. Other problems will occur randomly, such as a piece of debris clogging up a particle counter flow cell. In these cases flow alarms will usually signal the problem. If no alarms are available, a sudden drop in the data output is a warning sign. This is the reason particle counting system operation and maintenance functions should not be kept separated. The staff should have sufficient training and be allowed to gain experience with the system to be able to spot such problems quickly.

If a particular particle counter has a lot of "random" problems such as clogging or contaminant buildup, it should be evaluated closely. The installation should be revisited and any appropriate changes made. It may be necessary to add more frequent checks to the maintenance schedule. Unscheduled maintenance problems should only occur because they are unpredictable.

3. Maintenance Log

It is a good idea to keep a log of all scheduled and unscheduled maintenance activities. Over time, experience will become the best guide for setting maintenance schedules and minimizing unexpected problems. A written log will also aid in transferring that wealth of experience to new personnel. At the beginning of a new log, summarize the maintenance experiences from the previous period, to provide a baseline for comparison, and to keep that information ready at hand.

4. Maintenance Checklist

Compile a list of items to check for each particle counter. Make sure that all maintenance personnel are performing the same tasks, and doing so consistently. Brief guidelines for performing each operation should be listed, with a space for recording any measurements or observations. It may be a good idea to laminate copies of helpful diagrams from the particle counter operation and maintenance manual and keep them on the clipboard used for the checklist.

Information from each checklist should be recorded in the maintenance log. It should also be given to the system operator to reference in the data files. Any interruption in operation for cleaning or adjustments in sample flow will alter the data output from that particle counter, and must be accounted for.

5. Flow Maintenance

The first check performed should be to verify that the sample is flowing through the particle counter at the correct flow rate. This should be performed routinely even if a rotometer or electronic flowmeter is in use. The most reliable way to check the flow is with a graduated cylinder and a stopwatch. Collect at least 30 seconds worth of sample in a properly graduated cylinder. Do not use a 1-liter cylinder to measure a 100 ml/minute flow rate. A 100-ml unit is easier to handle, and more accurate. Take the sample from the outlet tubing of the particle counter. Be careful not to raise or lower the sample outlet as this will change the head pressure and alter the flow.

If the flow is only off by a couple of percent, adjust the sample outlet slightly by raising or lowering it to the proper height. Take another measurement to verify that the adjustment has been made properly.

If the flow is significantly lower than normal, some sort of obstruction is probably the cause. The following steps should be taken to correct the problem.

1. Verify that the constant-head overflow weir is working properly. An overflow should always be present. If no overflow is visible, check the flow into the constant-head overflow weir.
2. Remove the inlet tubing from the particle counter. A steady stream of sample should be flowing out of this tubing, well in excess of the required flow for the particle counter. If this in not the case, the problem exists in the constant-head overflow weir.
3. In most cases the obstruction will occur in the flow cell of the particle counter. If a flowmeter is installed, it could be clogged. Remove the tubing connecting the particle counter to the flowmeter. If the blockage is in the particle counter flow cell, the sample will trickle out. If the sample from the particle counter is streaming out, the flowmeter is clogged.

If the constant-head overflow weir is set up properly, the flow should not vary much at all unless the flow path is obstructed. One exception is the buildup of bubbles due to a change in temperature undergone by the sample. In some climates, the sample may start out only a few degrees above freezing. If sample lines are run into a heated area, bubbles will come out of solution, and reduce the flow through the particle counter. Bubbles can also affect flowmeters, which is the reason hands-on checks are necessary. Bubbles can also result from improper sample tap location, or entrained air in the process water. Bubbles may also be counted as particles.

6. Cleaning

Each brand of particle counter will have different recommendations for cleaning the flow cell. These are discussed in Part III of the book. Consult the particle counter operation and maintenance manual for each model before cleaning. This section will present only general guidelines.

Two types of cleaning should be distinguished. The first involves removing buildup from the flow cell windows, which obstructs the optical path, but not the flow path. Obstructions to the flow path are usually the result of debris becoming trapped in the flow path.

a. Coatings on Flow Cell Windows

All particle counters monitor the light energy produced by the laser as it passes through the sample. As the cell windows become coated, the amount of light measured decreases. At a certain point, the particle counter will output an alarm to indicate that the windows need to be cleaned. This buildup can occur slowly over several weeks, or rapidly in some cases, such as the addition of permanganate into the process.

Cell windows should be cleaned when indicated by the alarm on the particle counter. Systems may provide a numeric output, such as a 0 to 100% scale, to allow precise monitoring of the buildup. The numeric scale allows for proper maintenance to be performed before the alarm point is reached. Less-sophisticated particle counters provide an "idiot light," which comes on when the unit must be cleaned. These units require a more rigorous cleaning schedule. Reduce the time between cleanings until these alarms rarely occur.

Cell windows are usually cleaned with a very small brush, designed to fit snugly into the flow cell. A standard laboratory glassware cleaner is used as a cleaning agent. Iron or permanganate buildup can be cleaned with a slightly acidic cleaner, such as Hach Rust Remover.

The cell should be cleaned until the indicator returns to 100%, or until the light goes away. Again, the less sophisticated "idiot light" leaves the quality of the cleaning job subject to guesswork. *Note*: The cell indicator will not read accurately until the flow cell is full of sample.

b. Clogs and Flow Cell Obstruction

The second type of cell cleaning is the removal of obstructions from the flow path. These problems are usually cases of random debris or otherwise "unscheduled" maintenance. They may be indicated by cell alarms or low-flow alarms. Usually, the clog will occur at the entrance of the flow cell, where the flow path narrows down to 1 mm^2 or less.

Most clogs of this type can be removed by reversing the sample flow through the flow cell. Remove the inlet and outlet tubing, and connect the inlet tubing to the outlet port of the particle counter. If the debris is not lodged tightly, this should be sufficient to flush it out. If the debris will not flush out, try compressed air or a higher-pressure water source. *Note*: Do not exceed the pressure limits of the flow cell. Check your operation and maintenance manual for guidelines.

Do not use the cleaning brush to clear clogs from the flow cell. The brushes are designed to clean the middle of the cell windows, and do not extend all the way through the cell. Clogs usually lodge on the edge of the flow cell in the inlet port. They must be removed by pushing them out from the outlet port We have seem an immeasurable number of bent and broken brushes, which have been used in failed attempts to remove clogs. They are not terribly expensive, but are only available from the particle counter manufacturer, and are usually in short supply at the water plant.

A particularly nasty type of clog can occur when long-neglected sample valves are opened for the first time. If the valves are not flushed before the sample tubing is connected, a sticky blob of gunk can shoot straight into the particle counter flow cell, and it is nearly impossible to remove. Always flush out sample taps before connecting to the particle counter.

7. Maintaining Sample Tubing

Sample tubing should be checked regularly for excessive buildup, discoloration, or bubble formation. The standard Tygon ™ type tubing will develop a buildup of particles on the inner walls. This is due to the material properties of the tubing, and cannot be prevented. Tygon™ will also yellow from the chlorine in the water. There is no need to try to clean the tubing, as it is inexpensive and easy to replace. How often this will need to be done will vary from plant to plant.

Excessive particle buildup will result in particle "shedding," which is a periodic release of excess particles through the particle counter, resulting in erroneous counting. Sunlight striking clear tubing will produce organic growth, which will cause shedding and constrict the tubing. Be sure to check for seasonal variations and their effects on the tubing.

The only way to determine if the particle buildup in the tubing is causing problems is to make note of changes in the data output when the tubing is replaced. Allow time for the new tubing to be flushed out thoroughly, and compare the data with that taken previously. Make note of the results in the maintenance log.

Keep in mind that moving the tubing, or "flicking" it to clear particles or bubbles will cause an increase in counts. The particle counter is extremely sensitive to any change in the particles passing through. Always make note of any slight adjustments in the maintenance log, so the system operator can account for increases in the data.

The constant-head overflow weir may also require periodic cleaning if particle buildup occurs. Anything in contact with the sample is a source of contamination. It is impossible to prevent a certain amount of contamination, but reasonable attempts to minimize it will be necessary. Additional flowmeters or alarms should be maintained according to the procedures outlined in the operation and maintenance manual.

8. Strainers

Some particle counter manufacturers recommend strainers on the sample inlets to prevent clogging of the flow cell. In most cases, they are unnecessary, and may cause more trouble than they prevent. The only case where strainers make sense is on raw water sample lines, where fairly large debris can enter from the environment outside the plant. Once the water has been settled and filtered, particles large enough to clog the sensor should be rare. Strainers provide a great trap for particles that will shed periodically, skewing the count data.

If a strainer is used for raw water sample inlets, it should have a fairly large capacity to prevent constant clogging. The mesh openings should be just a slight bit smaller than the particle counter flow cell. The only reason for the strainer is to collect particles that can clog the flow cell, and there is no need to collect anything smaller, as that will only increase maintenance requirements and skew the data.

Some manufactures may still supply or recommend strainers for their systems. This could be because of the difficulty of accessing the flow cell for cleaning, or because someone thought it was a good idea several years ago, and no one ever questioned it. It would be a mistake to assume that good reasons exist for everything promoted in the particle counter industry, just as in all other areas of life.

The best recommendation is not to use strainers unless the operation and maintenance manual explicitly mandates them for a good reason (or one such as "removal of strainer will void the warranty"). If a sample point proves to be a problem, a strainer can always be added later. Be especially wary of putting strainers on settled water lines, as the flocs can coat them, quickly, and create a lot of problems. The velocity of the sample slows down through the strainer because of its large volume relative to the sample tube. This allows the floc to stick to the screen. The sample velocity through the particle counter flow cell is usually high enough to push these floc particles through without a problem.

9. Pilot Plants and Other Special Applications

Pilot plants can present special problems for sample collection and flow maintenance. These problems are usually the result of wide variations in pressure and flow due to the compact size of the pilot plant. Filter columns may not provide adequate head over the course of a filter run. Just finding the space to make sample taps and mount the particle counters can be a problem.

Problems in these installations are somewhat mitigated by the fact that only a few particle counters are typically needed in a pilot setup, making the use of pumps or other special equipment less impractical. There is usually more oversight of equipment in a pilot setup, so problems can be addressed more readily.

A turbidimeter with a built-in constant-head reservoir provides a good location for pulling a sample, as described above. If a pump is used to keep flow constant, it should be located downstream of the particle counter to prevent contamination. The sample should be drawn from a reservoir open to atmospheric pressure, to keep a constant inlet head pressure. This can be set up by running the sample from a pressurized tap into a small beaker that is small enough to allow the sample to overflow into a sink or floor drain. This will keep a constant inlet pressure, as well as flush out any particles falling into the beaker from the air.

Peristaltic pumps are a good choice for these applications. They will not jam up because the particles never contact the mechanical parts of the pump. They produce pulsations in the flow, but this can easily be corrected with a small pulse dampener. Several pump heads can be ganged together on a single drive, allowing for multiple samples to be controlled. External tubing clamps can be used to regulate flow. Some peristaltic pumps have variable-speed controls. Metering pumps provide a smoother flow, but can be jammed by large particles.

These are just a few suggestions for handling pilot plant installations. As long as the basic rules are kept in mind, any number of approaches can be tried until the most workable solution is found.

D. CALIBRATION

All instruments must be calibrated to ensure proper performance. Particle counters require a somewhat specialized calibration, which is described in detail in Chapter 14. Particle counters will hold calibration for over a year in most cases, because of the nature of the laser light source used. Because particle counter calibration is specialized and infrequent, it is best left up to the manufacturers or qualified third parties. Most provide on-site as well as factory calibration. This section will provide a brief overview of the calibration process, with the primary emphasis directed toward choosing the best approach for handling calibration from the standpoint of cost and efficiency.

1. Particle Counter Calibration

At present, particle counters are only calibrated for sizing accuracy, and not for counting accuracy. This calibration is performed by passing particles of known size through the particle counter sensor, and measuring the amplitudes of the pulses generated for each particle. This information is then used to adjust the counting electronics to distinguish among the sizes accurately. Usually 10 to 12 different sizes of particles are required for a complete calibration.

Needless to say, this process requires a good understanding of particle counter technology, and some accurate equipment for making measurements. While not excessively complex, it is beyond the capabilities of the typical treatment plant. An investment in equipment and training is required, which cannot be justified unless a large number of particle counters are installed in the treatment plant, or in a large municipal water system with several plants that can pool resources to develop a specialist in this area.

How many particle counters are required to make in-house calibration a practical alternative? That is difficult to say, because it depends a lot on the individual situation, but a good estimate would be around 50 or so. Since particle counters typically are calibrated only once a year, it takes quite a few just to stay in practice. For most plants, in-house calibration will never be practical.

2. Particle Counter Calibration Verification

Verifying calibration is an easier and more practical step for the water plant operator. Only a couple of particle sizes are required, and the goal is to determine if calibration is required. While less complex, this procedure requires attention to detail and a good grasp of how particle counters work. In most cases, someone with good laboratory skills would be the best candidate. The various approaches to particle counter calibration verification are discussed in Part II.

3. Maintaining Calibration

For the majority of readers, particle counter calibration will consist of scheduling for annual calibration with the manufacturer or other qualified outside source. Two alternatives are available. The units can be calibrated onsite, or returned to the factory or regional service center. Which one is chosen will be dependent on cost. While on site calibration is more desirable for obvious reasons, it is not always cost-effective. Travel costs must be factored in, and amortized over several units.

If the calibration is to be done on site, then all the particle counters should be calibrated at the same time. If other plants in the area own units made by the same manufacturer, the calibrations can be scheduled for the same trip, allowing travel costs to be shared, and making on-site calibration more practical for small plants.

If the particle counters are to be returned to the factory, then calibrations should be scheduled sequentially, so that only one unit is out of service at a time. If a spare is available it can be rotated into service. The maintenance log should be kept up-to-date, and calibration stickers displaying the date of calibration placed on each particle counter.

CHAPTER 4

Collecting Data

Operation and maintenance procedures are essential for a workable particle counting system. They are what filling the gas tanks, checking the oil and tires, and packing the car are to taking a trip. They are indispensable, yet only bring us to the beginning point. The success of the journey depends upon foresight, planning, imagination, and the ability to deal with unexpected events along the way. The same holds true for operating a particle counting system. Just as a journey to a new place presents the traveler with a wealth of sights, sounds, and smells which form the experiences that make the trip worthwhile, the particle counting system will provide a vast amount of information. Whether that information is received as an overwhelming amount of confusing and useless data or as a wealth of valuable clues waiting to be pieced together to provide new avenues of learning and experience will depend a lot on the individual operator. Whether one takes a series of unexpected and challenging occurrences on a trip to be frustrating hassles or exciting adventures depends on the makeup of the traveler. Proper training and information are necessary parts of that makeup, and providing them the goal of this book. The more essential elements of character, integrity, and desire to do one's best are left to the reader.

These points are brought up because particle counting is still pretty much an unregulated technology in the drinking water industry. This allows for particle counting systems to be used to great advantage by the innovative and adventuresome, and all but ignored by those unwilling to commit the time and effort required. The rapid growth of particle counter technology in the drinking water industry despite the lack of regulations requiring it speaks well for its usefulness. It also points out the debilitating effects of overregulation, which can turn a minimally acceptable standard into "the" standard, and can kill off innovation. One reason that particle counters have provided such an advance in a short time period is that turbidity regulations ended real innovation in that technology many years ago. Overregulation results in time and resources being committed inefficiently and takes them away from areas more profitable. The infamous "Lead and Copper Rule" is a case in point.

Particle counting will not be regulated for quite a while, thanks to many unresolved issues that are explained later in this book, and the fact that the regulatory agencies are too busy regulating a host of other areas of life to get around to it yet. The one state which has mandated particle counting for most of its plants requires a minimal amount of data to be reported. It is obvious that its intent is to get the plants to use the technology to improve water quality without becoming burdensome. This prudent form of "encouragement" was instituted by an agency that works well with its plants and is knowledgeable about particle counting, as a response to one of the first known *Cryptosporidium* outbreaks in the country. Its goal is to improve overall plant performance to keep ahead of ever tightening EPA standards. Particle counting is considered one of the most cost-effective ways to achieve this goal.

This chapter discusses the approach to handling the data produced by the particle counter system, and how to use the data as diagnostic tool and not just a checklist. We will attempt to provide a framework for developing a useful and practical approach to particle counting that can be tailored to the capabilities and resources of the individual drinking water treatment plant. None of the information presented here is intended to be specific to any brand or product. Those details are presented in Part III of the book.

A. DATA COLLECTION

Before the data generated by the particle counting system can be interpreted, it must be collected, displayed, and stored. Water operators are familiar with circular and strip chart recorders, which have been a commonplace for many years. Although perhaps not obsolete, they are not practical for particle counting in most situations. The only practical choice is the personal computer. There is one argument that can be made against this option, and that is that some operators are not familiar with computers. The answer is that it is time they learned. Like it or not, the computer will be the centerpiece of drinking water treatment plant data collection from now on. Consider the reasons:

1. A single personal computer can collect display and store all the data produced by a large water treatment plant, and costs less than a single chart recorder. (Taking into account the software and initial setup costs involved, it may cost as much as a dozen chart recorders, which is still a bargain.)
2. Data collected on a personal computer can be reorganized readily to allow comparisons between different instruments or time periods, and can be processed statistically. Chart recordings are frozen in form, and require manual manipulation for any type of comparison.
3. Historical data can be retrieved and stored easily, backed up in several locations, and transported at virtually no cost.
4. The great improvements in speed and processing power of the personal computer has resulted in simplified graphic user interfaces (GUIs) which make basic operations as easy as playing a video game.

5. The huge increase in regulations has made larger amounts of data collection and reporting mandatory while the costs of compliance have driven plants to reduce manpower. These trends have made economizing data collection a necessity. The personal computer provides the only practical way to achieve this.

If all these reasons are not enough to be convincing, it will become obvious as one reads on that particle counters are especially suited to computerized data collection. To list all the reasons this is so would be redundant. A brief example should serve to illustrate this point. Suppose a single particle counter is used to provide counts in four size ranges. An alarm for low flow and one to indicate a dirty flow cell are also used. Now add in the filter removal efficiency calculations (log removals) which require the data from another particle counter to be combined and calculated with that of our example. Now, 10 data points (including calculations) are now being produced for a single particle counter. Need we say more?

B. DATA PRESENTATION

The way the particle counter data is presented will bear directly on the usefulness of the data. The mass of information produced by the particle counters will easily overwhelm the operator if it is not organized and displayed in a logical manner.

The organizing principle must be based on the primary purpose for particle counting in drinking water treatment. That purpose is the optimization of particle removal throughout the entire course of each filter run. Once this purpose is understood, the organizational structure for presenting the data will become clearer.

1. Trend Display

The most important and useful data presentation is the trend display of particle counts and removal efficiencies over time. The goal of every drinking water treatment plant is the consistent production of high-quality drinking water. Trends provide a direct way to monitor this continuous process quickly and accurately. Trends put individual data points into context, and provide a sensible framework for interpreting unexpected changes in that data.

The importance of trend displays to understanding and using particle counters cannot be emphasized enough. Most of the uncertainty and confusion surrounding particle counting would be eliminated if operators would work with trends and quit being overly concerned about individual count values, or inexact count correlation between individual particle counters. Although these issues are important, and are dealt with in later chapters, they should not prevent the drinking water operator from gaining a wealth of knowledge and valuable information from particle counting systems as they exist at present. Much of this reaction is understandably the product of a regulation-obsessed industry, but it is no excuse for not using particle counters to great advantage.

2. Trend Particle Counters with Other Plant Data

To relate the particle counter to overall plant performance, trend it together with other plant data. Turbidity, headloss, and any other data that relate to water quality and filter performance can be used to provide the background for interpreting particle counter data. Once again, the trend presentation provides the most complete and easy-to-understand picture of the interrelationship of these parameters.

To facilitate this, the data presentation should provide the means for selecting individual parameters for simultaneous trend display. Most likely, this will be done with computer software, and requires that the plant instrumentation data be available to the particle counting system, or vice versa. The various ways this can be accomplished are presented below. For now, we are concerned with the data presentation itself.

It is also important to compare the particle count data from different particle counters on the same trend display. This allows for comparing filter effluents, or influent vs. effluent. Ideally, trending multiple data points together should be quickly and easily achieved. In most cases, four trends per display are sufficient. Too many trends crowd the display, making it difficult to follow the individual parameters. The data are trended over time, and the display should allow selection of time spans that provide good resolution, while permitting a full filter run to be displayed when desired.

3. Other Data Displays

The data may also be presented in tables or other numeric displays, which are secondary in importance to the trend displays. Numeric presentations are useful for determining exact values, which may not be discernible from the trends. They are necessary for daily averages or maximums and minimums, and for reporting purposes. In most cases, numeric tables are not helpful for interpretation, and should not be emphasized as such.

4. Data Reporting

Until particle counting is mandated by regulations, most reports will be generated for internal records and for review in cases of problem occurrences. They should be set up in a way that provides an efficient yet sufficient presentation of the data for the period covered. In most cases, they will not be referenced unless a problem has occurred. The report should contain sufficient data to point out any odd occurrences or problems encountered with the particle counting system. Once the particle counting system has been in use for awhile, the operator should have a good grasp of normal operating conditions, and can tailor the reports to this end.

5. Historical Data

The lower costs of computer equipment have resulted in lower costs for data storage. Historical data should be maintained as long as is practical, which should be for a couple of years or more. All the data do not need to be kept, but enough to show any odd or unusual occurrence. Again, that determination will have to be made by the operator once sufficient experience has been acquired.

Storage concerns are not as important as the ease of retrieval. The data should be logically organized so that information for a given time period can be located and displayed readily. This will make accessing historical data practical for more than just emergency situations, such as reviewing seasonal effects from year to year.

C. SYSTEM STRUCTURE

Particle counting systems are usually built around some type of computerized data collection. The data collection computer will be either a "stand-alone" unit provided by the particle counter manufacturer or the plant data collection computer (known by various acronyms as SCADA, DCS, etc.). Sometimes a combination of the two is used. In a few cases, chart recorders are still in use, and require the familiar 4 to 20 mA current loop signal output. This section is meant to provide a broad overview of the available options, and the rationale behind each. Technical information for these approaches are covered in Parts II and III of the book.

Determining the best approach for a given plant application requires balancing the costs and operational efficiency with the primary goal of particle counting for drinking water treatment; that is, continual optimization of water quality. It is also necessary to gauge the commitment of plant management and of the operations and maintenance staff to particle counting. If minimal interest is shown in the technology, then the simplest, lowest-cost approach is probably the best. If an initial trial with a couple of particle counters has proven invaluable, and particle counting is to become a large-scale, integral part of the treatment system, a different approach is required. These determinations must be made with reference to each case, and are beyond the scope of this book. Some general guidelines will be presented, which should help in this decision process. We will begin with the most straightforward, in terms of initial setup and operation.

1. Turnkey System

By "turnkey," we mean simply a complete system provided by the particle counter manufacturer. This includes the computer and software for data collection, as well as all the particle counter equipment. It is not necessarily a complete installation, as the plant personnel may run signal wiring and mount the particle counter hardware. The computer may even come from a separate source. But basically, the system comes complete from the manufacturer. It involves a standardized setup without reference to the existing plant equipment.

From a cost standpoint, it is straightforward to estimate the number of particle counters as well as the computer and software. The only variables are cabling lengths and installation. The manufacturer can provide full support because a standard package is easy to document and install. In all cases, it is a good idea to get standard, turnkey system pricing for any new particle counting system installation, as it provides a baseline for comparing other, more complicated, approaches.

The disadvantage to this approach is that the particle counter data are collected and displayed separately from the other plant data, requiring separate training for the operators, and making comparative analysis of particle counts to other plant data difficult. This tends to isolate particle counting from the rest of the plant operation. Often, only one or two operators become proficient with the particle counting system, and the rest gain little or no benefit.

2. Turnkey System with Additional Inputs

To provide the benefits of trending particle counts with other plant data, most of the manufacturer's turnkey systems provide additional signal inputs for plant instrumentation. These are usually in the form of current loop or voltage inputs. Some also provide contact closure (discrete) inputs for signaling the backwash valve position or other events that can impact the particle data.

Again, the systems are not truly turnkey in that the plant personnel are usually responsible for connecting the auxiliary signals into the particle counting system. In most cases, the particle counter equipment is being added in long after the plant instrumentation. The current loop signals from these existing instruments must be rerouted to the appropriate inputs on the particle counting system without disturbing the existing signals being sent to chart recorders or the plant SCADA system. The particle counter manufacturers are not equipped for this task, and do not want to risk damaging the plant instrumentation system, so installation must be performed by the plant electrician or an outside contractor.

In most cases, only a few of the plant instruments will be tied into the particle counting system. Turbidity, headloss, and streaming current are often trended along with particle counts. This helps provide a context to make the particle count data more understandable, but still leaves the problem of maintaining two systems.

3. Particle Counters Tied Directly to the Plant SCADA System

This approach is the most appealing in terms of making the particle counter data an integral part of the plant operational system. All data collection, display, historical data storage, and reporting are kept on a single system, providing maximum utility for the operator.

Unfortunately, this is the most-complicated way to set up the particle counting system. Several different ways of approaching this difficult task have been attempted, and are summarized below:

a. Particle Counters Integrated Directly into SCADA

This involves connecting the individual particle counter units directly to the plant data acquisition system. Two methods are available. The first involves particle counters with 4 to 20 mA current loop outputs. This is relatively simple from the SCADA standpoint, as SCADA systems are designed to receive current loop signals. The problems in this case are inherent in the particle counters, which are not well suited to 4 to 20 mA current output. A whole range of issues pertaining to this will be discussed in a special section below.

The second alternative is the direct interface of the particle counters to SCADA using serial data communications. All count data, as well as status and alarm information, are transmitted in a specially coded format over a twisted-pair data line. This is a fairly complex procedure, and is well beyond the scope of this book to explain in its entirety. Some technical details are presented in Parts II and III. The important thing for the plant operator or consulting engineer to understand is that these details should be covered thoroughly before a direct particle-counter-to-SCADA-system integration is specified and purchased. The particle counter manufacturers and competent systems integrators should be able to work together to accomplish this task. *Be forewarned*: The "We'll work that out after the bid" type of approach is a less than intelligent one, to put it nicely.

A brief analogy should help make the problem more understandable. Serial data are a combination of thousands of "on" and "off" signals grouped into "bytes" and transmitted according to some predefined "protocol." If one thinks of these on and off signals as various sounds produced by human speech, the bytes would be analogous to words, which are composed of various combinations of these sounds. The combination of sounds into words and words into phrases would be ordered according to the grammar of the particular language (or protocol) involved. Each device is designed to communicate via a certain protocol, or language.

Let us assume that the SCADA system "speaks" French, and the particle counter "speaks" Russian. How could the two be made to communicate? Obviously, some sort of translator would be required. This is usually referred to as a "driver," and is a special type of software that "translates" the data received from the particle counter to the protocol of the SCADA system.

Many different makes of SCADA software are available, each with its own protocol. A large number of instruments designed for digital serial communications are also available. Widely used instruments have had special drivers developed by most of the SCADA system vendors. These are usually available for purchase from the SCADA supplier. It is sort of like the United Nations, where one can hear a speaker translated from almost any language into almost any language. Unfortunately, most of the particle counters available have unique protocols, which are not provided for by the SCADA system vendors or third-party suppliers. They are akin to a UN delegate from a newly discovered tribe, with a language no one else knows or understands.

Many instruments have been designed to take advantage of the popular protocols supported by the major SCADA packages. Unfortunately, with one or two exceptions, the particle counter manufacturers have not taken any steps in this direction. It has been said that generals are always preparing for the last war, and, when it comes to streamlining for SCADA interface, most of the particle counter manufacturers seem to be awaiting the invention of gunpowder. Whatever the reasons, their inaction leaves a lot of work for the end user.

In time, drivers will become available as demand rises. Drivers can be written by third-party vendors for a particular application. If the particle counters are to be integrated into an existing SCADA package, options are obviously limited. If both the SCADA system and the particle counters are to be purchased together, some advance planning can pay off. The SCADA supplier may be enticed to create a driver for the particle counters if that closes the sale. The particle counter manufacturers may help cover the development costs for the same reasons.

Some particle counter manufacturers have developed drivers for specific SCADA packages. Often they provided these SCADA packages with their systems before developing their own software packages (software development is another area where most were dragged kicking and screaming). It is worthwhile to investigate these drivers, but be warned that some of them do not work well.

The large amount of data produced by the particle counters should not be forgotten when investigating SCADA integration. Make sure the SCADA package has adequate capacity for dealing with such large amounts of data. It should also have the flexibility to present the data in useful trend formats as described above.

It is not our intention to scare away anyone wanting to integrate particle counters into the plant SCADA system. This is the best option if it can be realized with reasonable effort. As time goes on, it will become less problematic. All the trends are pointing to this option as the course of the future. Until then, intelligent planning and forethought are the keys to avoiding an ugly, expensive mess.

b. Hybrid Approaches

Several options exist for combining the turnkey system with SCADA integration. One has already been mentioned, that of sharing the plant instrumentation 4 to 20 mA signals between the SCADA and the particle counting system. It is also possible to send data from the particle counting system to the SCADA via file sharing over a computer network. In this arrangement, the particle counting computer can send relevant information to the SCADA system to be trended along with the plant data. Integration costs may be lower than that of the direct approach, since the SCADA system does not have to process all the particle count data. File sharing may not require a special driver interface. There is still the added complexity of employing two systems, but since the data are available to all the operators on the main SCADA system, it is not a problem to have one specialist maintain the particle counting system. The particle counter manufacturers are more amenable to this approach, since it requires less effort on their part than direct integration.

A less desirable but simpler option is to run the turnkey particle counting system, while taking an additional 4 to 20 mA signal out of each particle counter to be input into the SCADA system. Section 4 below covers the problems with 4 to 20 mA outputs. One immediate problem is that the 4 to 20 mA data will not match the digital output data from the particle counter. The reasons for this are easy enough to understand. From an operator's standpoint, different readings can undermine confidence in the system.

It is also possible to send data from the SCADA system to the particle counting system, although it is much less likely. SCADA software tends to be more flexible than that designed for particle counters. It is doubtful that such an approach will be practical.

4. 4 to 20 mA Current Loops

As promised, a special section is devoted to 4 to 20 mA current loop outputs. Whether the particle counters are interfaced to the SCADA system, a data-logger, or a chart recorder, the problems are the same. The 4 to 20 mA current loops have inherent limitations and errors regardless of the instrument producing them. The special nature of the particle counter further compounds these problems.

a. Digital vs. Analog

To understand the inherent limitations of the 4 to 20 mA current loop, let us begin by examining the differences between analog and digital data transmission. All data transmission is a transfer of information. Analog data are transmitted exactly as they occur, and anything that affects the method of transmission will also affect the data. A vinyl phonograph recording is an example of analog data transmission. Each track on a phonograph record is a continuous groove that runs around the record, and varies in shape exactly as the recorded sound varies. Any scratch on the record alters the recorded sound by causing a pop or static sound. A certain amount of "hiss" can be heard in the background. If the speed of the turntable varies, or the phonographic needle is damaged, the sound heard will not be a true reflection of the recorded sound.

Digital data transmission is used to prevent transmission-related problems from affecting the information being transmitted. The information is converted from an analog signal into a digital format by means of high-speed electronic circuitry. Digital information can be transmitted without error because it consists of a pattern of on and off or high and low signal levels. Whereas analog signals are continuously varying signals of virtually an infinite number of levels, digital signals have only two levels, spaced widely apart. The data being transmitted are not affected by outside interference or the quality of the receiving equipment. An analogy could be made to listening to a violin sonata vs. listening to a string of gunshots. Those with various degrees of hearing impairment, or who are located next to a noisy patron, will not receive all the information transmitted by the violinist. Few will miss the gunshot sounds.

The popular compact disc recordings, which have replaced phonograph records, are examples of digitally transmitted data. Scratches that would ruin an LP will not alter the sound from a CD. The speed at which the CD is played is not critical, because time spacing information is encoded along with the sound, eliminating a source of mechanical error. In short, the CD provides a great advance in sound transmission technology because transmission-related errors have practically been eliminated.

Nothing is perfect, and digital data transmission has its own problems. The trade-off for eliminating the transmission problems associated with analog signals is that a lot more information must be transmitted to get the same amount of data across. In this sense, digital data transmission is much less efficient, and requires more complex and expensive technology. Transmission problems are related to keeping all the data in the right place and transmitting the data quickly enough. Fortunately, the nature of digital data is such that its accuracy can be verified by the receiving equipment. The data are transmitted along with a derived value called a "checksum." The checksum is also calculated by the receiving equipment and matched up with the transmitted checksum to verify that all the data have been received properly. It is sort of analogous to counting school children on a field trip. If the numbers do not add up, you return to the last point and start over. Digital information can be retransmitted because it includes its own time information. Analog information is transmitted in "real time" and cannot be recovered if not received correctly the first time.

b. Specific Sources of Error in 4 to 20 mA Current Loops

Several sources of error are inherent in 4 to 20 mA current loops. Offset errors can result from improper calibration of the transmitter or receiver. Noise can be induced from external equipment. The signal resolution is only as good as that of the worst component in the loop, whether the transmitter or receiver. The data transmission is in one direction only. The transmitter cannot "know" if the data it is sending are being received, and the receiver has no way to verify if the data it is receiving are correct.

c. Special 4 to 20 mA Problems in Particle Counting

In truth, 4 to 20 mA current loops are especially ill suited to particle counting. The first and most obvious reason is that each particle counter produces several channels of data. The cost and complexity of sending three or four current loops out of each particle counter becomes excessive if more than two or three units are involved. At least one alarm should be transmitted to signal instrument problems, so a four-unit particle counting system could easily require 20 or more outputs. With proper isolation, the cost of a single analog input to a SCADA system can easily reach $200 per point. This small four-unit system would require enough additional hardware to cover the cost of a good personal computer and particle counting software program. The same holds true for chart recorders, as discussed above. This initial cost is in addition to the calibration and maintenance requirements for all these signal transmitters and receivers.

Beyond these practical considerations are the problems resulting from the limited resolution inherent in 4 to 20 mA systems. Most 4 to 20 mA receivers can resolve the signals on their inputs at 10 bits of resolution. This means that they can separate the signal into 1024 distinct values. (This is the way analog data is broken down so that it can be converted to a digital number to be processed by the SCADA system — somewhat ironically.) Resolution can also refer to the width of the pen on a chart recorder. Changes in signal less than this cannot be distinguished.

Why is this important? Let us use turbidity as a comparison. Consider a filter effluent. Current regulations require that turbidity be kept below 0.5 NTU. Good operating guidelines suggest that filter effluent turbidity be kept below 0.1 NTU. If the span of the 4 to 20 mA signal is set from 0 to 1 NTU, this range is covered nicely. With 10 bits of resolution (1024 steps), turbidity can be resolved down to 0.001 NTU. This is a full order of magnitude better than a turbidimeter can reliably measure.

Now let us place a particle counter on the same filter effluent source. It is possible to count particles well below one per ml (remember that a 25-ml sample volume is typical) up to several thousand per milliliter on water that measures less than 0.5 NTU. A 10-bit system would allow us to span from 1 to 1024 particles/ml. This is about one order of magnitude less than the range of particle concentrations measurable below 1 NTU. Consider that the best 4 to 20 mA receivers are capable of 12-bit resolution (4096 steps). This is still less the half the measurable range. Some older receivers have only 8 bits (256 steps) of resolution. The reader can work out the math for that one.

Obviously, this poor resolution makes 4 to 20 mA current loops a less-than-desirable option. It is true that a properly operated plant will produce counts less than 100/ml in many cases. This still places the median operating point in the bottom 10% of the signal span. Spikes over 1024 particles/ml will be off scale as it is, so shifting that median up to the center of the span is out of the question. The scale can be compressed to read every second or fourth particle, at the cost of sensitivity. This will have to be done for settled or raw water sources, but it is a pity to reduce the sensitivity of the particle counter when high sensitivity is one of the major advantages of the technology. Spikes may not occur often, but they are important occurrences that need to be captured as accurately as possible.

It is important to keep in mind that these resolution limitations are not taking into account any of the other sources of error mentioned above. Whereas our turbidimeter readings have a full order of magnitude of excess resolution, we are already losing ground with the particle counter before these errors are considered. These errors are further exaggerated when log removal calculations are performed.

The point should be sufficiently made that using 4 to 20 mA signals for particle counting is a good example of the "tail wagging the dog." This is not to say that they should never be used, and each case will have to be examined on its merits. There are fewer and fewer "good" reasons for pursuing this option, and these will be further diminished as digital data interfaces become easier to integrate. If consideration of all the concerns outlined above is not enough to sway the reader from this approach, then it is probably a good course of action for the given situation. But we all know of cases where the path of least resistance was followed, and the resulting mess was blamed on the technology, not the poor preparation and planning. The equipment then went unused, and the plant operators were deprived of a useful tool, while the ever-impoverished taxpayer has dropped a few more ducats down the drain. There are, of course, cases where the technology itself is the problem. But particle counters have been proved to work well when implemented properly.

CHAPTER 5

Grab Sampling

Grab sampling is a familiar form of sample measurement. Among its advantages are lowered equipment costs and the ability to measure samples from any accessible place in the process stream. Some of the disadvantages are that more direct labor is involved, and the sample can be altered during collection and testing. Obviously, a much less complete picture of process changes can be produced in comparison to a continuous online measurement. While this is true for all forms of grab-sample data collection, the high sensitivity of particle counters makes careful consideration of these factors imperative.

This chapter begins with a brief description of how grab-sample particle counters operate, then (as with 4 to 20 mA current loops) tries to talk the reader out of using them. After the reader has run this gauntlet and remained unconvinced, the discussion will turn to the practical aspects of grab-sampler operation.

A. PARTICLE COUNTER GRAB-SAMPLER OPERATING PRINCIPLES

A grab-sample particle counter is similar to an online particle counter in basic operation. The main difference is that an automated method of propelling the sample through the particle counter sensor has been incorporated, usually in the form of a pump. Unlike a turbidity grab sample, which can be measured directly from a stationary sample, the particle counter sensor operates at a fixed flow rate. This is because particle counts are measured per unit volume. The astute reader may point out that a fixed volume can be passed through the particle counter regardless of the consistency of the flow rate. This is correct, and some of the older, pressurized particle counter "batch" samplers operated under this principle. They were designed to dispense a fixed volume of liquid for each test run. Some samplers of this type are still in use in water plants. These units were designed to handle viscous fluids, which require a good deal of pressure to force the sample through the particle counter flow cell. They work well for water, but are much too costly. The manufacturers have found that adding a small pump to pull the sample through the flow cell works quite well for water. These units are designed to provide a constant flow rate for a fixed time period.

It should be obvious that the pump is used to pull the sample through the particle counter to prevent contamination of the sample. A sample is never run more than once, as it will be contaminated when it passes through the grab sampler.

Other than the pump, the only addition to the online particle counter necessary to make a grab sampler is the operator interface and data presentation. This can be done via a keypad and display, printer, personal computer interface, or a combination of any of the above. These options are covered in Parts II and III.

B. GRAB-SAMPLE PARTICLE COUNTING VS. ONLINE COUNTING

Since grab-sample particle counters are modified versions of the online variety, the major differences are to be found in the practical aspects of operation. Since the bulk of the book is devoted to online particle counting, it will be more efficient to point out the ways in which grab sampling alters the approach to particle counter application. There are many ways in which grab samplers make particle counting more difficult to apply, and a few areas where they are advantageous. Again, these observations are designed to provide the engineer or operations manager with the information necessary to select the best approach to particle counting for a given application. They are by no means the last word on the subject.

1. Reasons for Choosing Grab Samplers Over Online Particle Counters

In most cases, the reason for using a grab-sample particle counter is lower cost. Obviously, one unit costs less than an entire system. Akin to this is the intention to "start small" and determine how useful particle counting is before sinking a lot of money into a full-blown system. Both of these considerations are valid, and not to be dismissed lightly.

However, both of these considerations are becoming less valid as particle counting technology becomes more and more prominent in drinking water treatment. Costs have dropped a great deal in the few short years since particle counters were introduced to the industry on a wide scale. The same goes for the utility of particle counting. At some point, the usefulness of a given technology will become accepted without the need for each user to test its validity personally. When that point has been reached, the proper question becomes, "In what form will this technology be most practical for my given situation?" The ability of the plant personnel to accommodate particle counting should be the issue under consideration.

When grab-sample particle counting is viewed in these terms, it should become clear that "starting small" is not the same thing as "keeping things simple." Grab sampling requires much more operator involvement and attention to detail than an online system. If this is not understood from the beginning, and a grab sampler is purchased just to "try out" particle counting on a "small scale," the end result is likely to be unsatisfactory. In most cases, particle counting will be considered as something to experiment with after the mandatory tasks of the day have been completed. The grab-sample particle counter will require a substantial amount of time and effort to produce a significant picture of overall plant performance. As this becomes more obvious to the beginner, the result will often be that particle counting will become something to push farther back on the priority list. Or the plant super- intendent may not understand these problems, and wonder why particle counting is taking up so much of an operator's time. In many cases, the whole experience will lead to a distaste for particle counting, depriving the plant of the tremendous value this technology can provide.

A grab-sample particle counter should only be purchased as an introduction to particle counting if a highly competent laboratory technician will be operating the equipment, and is provided the necessary time each day to work with it. Otherwise, even an online particle counter with 4 to 20 mA outputs run into a chart recorder would be preferable.

2. Drawbacks to Grab-Sample Particle Counting

Two major drawbacks make grab-sample particle counting a poor proposition for most applications. The first is that each sample must be handled by an operator. The second is that only a sketchy picture of the plant performance can be achieved at best. Let us examine these considerations in detail.

a. Sample Handling

Particle counters are extremely sensitive to sample contamination. Their primary value is their high sensitivity to small quantities of microscopic particles. A review of the application data in Chapter 2 should make it clear that particle counters are much more sensitive than turbidimeters. Turbidity data are commonly collected from grab samples with little trouble. But 2 to 5 µm particles make little impact on turbidity data unless they are present in large quantities. It is hard to understand how easily particle count samples are contaminated unless one has experience in handling them. Keep in mind that a 2 µm particle is about 40 times smaller than what is visible to the naked eye.

Every step of the sample collection process is a potential source of contamination. Containers should be made of glass and washed thoroughly. Before the sample is collected, the sample container must be rinsed thoroughly with the sample. Sample taps should be flushed before dispensing the sample. The sample should not be stored for long because of settling, and a temperature increase can result in bubble formation.

Developing good sample-handling habits requires practice and attention to detail. No one works around the clock 7 days a week, so several operators must learn to collect and run particle samples. Any variation in methodology or different degrees of diligence in observing the precautions of sample handling can result in unrepre- sentative data.

Achieving consistent results in any science requires the elimination of as many variables as possible. Particle count grab sampling opens up a whole range of variables, which can only leave the validity of the data somewhat in doubt. This is especially true when an unexpected increase in particle counts is discovered in a filter effluent sample. Precisely the time that the particle counter is most valuable will be the time its reliability is most in doubt.

b. Grab Sampling Presents a Partial Picture

Not only do unexpected results cause more doubts about grab-sample data, in many cases important events will be missed altogether. Sampling several points in the treatment process more than once every few hours requires an almost Herculean effort. Covering a 24-hour shift will require at least three operators. The second and third shifts are usually minimally staffed, leaving little time for labor-intensive grab sampling.

While any particle counter data are better than nothing, a sketchy picture of overall performance is the best that can be achieved with a grab sampler. As long as things are working as they should, this is acceptable. But it is precisely the ability to detect potential problems before they become major problems that makes particle counters so valuable. Grab sampling does not preclude this benefit, but greatly reduces the odds of detecting problems at an early stage.

c. Data Handling

The nature of grab-sample particle counting makes data handling cumbersome. It has already been shown that particle counters produce a lot more data than other instruments in the plant. This data must be organized into some useful format. Since grab sampling is so unstructured, there is no easy way to automate data handling. It is usually left up to the operators to organize the data using a spreadsheet program. Log removal calculations will have to be performed manually for each effluent sample.

In summary, it should be clear that grab sampling is not at all "starting small" when it comes to particle counting. It is certainly true that some plants have developed useful and workable systems over time. That it takes time and commitment is the key. It is kind of like learning to swim by being thrown into the lake. You may learn, but it is certainly not the most efficient or desirable method.

3. Benefits of Grab Samplers

Now that the downside of grab-sample particle counting has been presented, it is time to explore the beneficial uses. There is one case where grab-samples prove to be easier to use than online particle counters. That is for high-concentration raw water sources. This is because dilution is straightforward with a grab sampler. The larger question of the usefulness of performing particle counting on conventional treatment raw water is still under debate.

Grab samplers are useful as a supplement to the online particle counting system. They can be used to verify the counts from each online particle counter, providing a ready standard of comparison. The grab-sampler unit can be sent out for a factory calibration, then used to check the calibration status of the online particle counters in the plant. Some grab samplers are capable of operating as online units, and can be used as a spare, or to monitor a point in the process where no online particle counters have been installed.

Since online particle counters are less susceptible to sample contamination, they can provide a baseline for developing accuracy with the grab sampler. This is a much better argument for making the grab sampler the last particle counter purchased rather than the first.

4. Alternatives to Grab Sampling

Since we have disparaged grab samplers to some degree, it is only fair that some alternatives for a low-cost way to "start small" be presented. The best recommendation is to start with a couple of online particle counters and a computer. This can be a basic turnkey system as described above. These systems provide a much better way to get an introduction to particle counting than grab sampling. Install one particle counter on the settled or applied (influent) source, and mount the other unit on a sawhorse that can be moved easily from filter to filter. Move it to a new filter every 2 weeks or so, to allow a couple of full filter runs to be monitored. Once it has passed through the full complement of filters a time or two, you will have learned a great deal about your filter operation, as well as the value of particle counting.

This "starter system" approach will give you a good taste of the full value of an online particle counting system, and costs little more than a single grab-sampler unit. Additional particle counter units are easy to add in over time. Chances are that your particle counting experience will be much better from the start, and the operators will not be soured on it before discovering the value it can provide.

If a SCADA system is already in place, a couple of 4 to 20 mA units can be tied into it to provide a good introduction. While not recommended for the full plant, one or two units are manageable. In most cases, the manufacturer can be induced to take them back in a "trade-up" to a full digital system.

These approaches will not afford the complete coverage of a full-blown system, but will get you off and running. The investment is small enough that not too much is lost even if the particle counters are replaced by a different make or model when a full-scale system is installed.

The intent is to make the initial particle counting experience as useful and pain-free as possible. That old maxim that "the first impression is the strongest" holds true here as well.

C. GRAB-SAMPLER SAMPLE HANDLING

Despite the strong warnings against it, grab-sample particle counting is not an impossible task. The requisite skills begin with a thorough knowledge of laboratory sample-handling techniques. Those not sure of what is meant here should skip to the next chapter. In other words, don't expect an operator who has no experience in the laboratory to pick up grab-sample particle counting and run with it.

Many of the problems inherent in collecting and running samples have been touched upon earlier in this chapter. To summarize them, the high sensitivity of the particle counter to particles of microscopic size means that the possibility of contamination is greatly magnified. The first caveat is that particle counters with a sensitivity of less than 2 µm are not recommended for grab sampling. Most typical plant applications do not require anything less than 2 µm, so this is usually not a problem.

1. Sample Preparation

Sample containers should be made of glass or Pyrex if possible. They are less prone to particle contamination than plastics, and can be inspected more readily for visible contamination. They should be cleaned and rinsed thoroughly prior to each use. Certainly an acid washer and particle-free storage area would be ideal, but these are usually not necessary for 2-µm particle counting. It is a good idea to keep the particle counter glassware set aside for only that use, to minimize problems.

Thoroughly rinse the sample beaker in the sample before completing sample collection. "Particle free" water can be produced with special filters, and is good for final rinses before the sample rinse. *Note*: Store-bought deionized water, or in-house-produced deionized water is not particle free. This will become obvious with experience.

Obviously, the most critical samples are filtered and finished water. These can produce less than 10 particles/ml at 2 µm in some cases, so contamination is of primary concern. Settled and raw water samples will usually be of high enough concentration to cover up any slight amount of contamination.

When collecting the sample, allow the tap to flush for a few seconds, then rinse the beaker thoroughly. Fill the beaker completely and let 50 ml or so spill over the top to flush out any particles around the brim. Pour off the excess, and then run the test as soon as possible.

Lids are not advisable, as they can be a source of particle contamination. If they are required to transport the sample, the lids should be cleaned and rinsed in the same manner as the container. Never use lids with coated paper liners, as they will produce a load of particles.

2. Sample Storage and Shipping

In cases where the samples are collected for in-house testing, they should be run as soon as practically possible. Bubbles will some out of the solution as the sample temperature rises, resulting in false counts. As for shipping samples to an off-site location, the best advice is not to do it. Particle counters are not nearly as expensive as when they were first introduced to the industry. The cost of having a sample run by an outside source is significant, so the number of samples that can be shipped and tested is small. This method is of little or no operational value, as results are not learned for several days, the possibility of contamination is great, and so few samples are tested that the results are virtually meaningless.

A good way to learn the effects of sample storage is to collect several samples from the same source, and test them several minutes apart. Compare the data for consistency. Consistency of results will be the best guide to determining the quality of sampling techniques in all areas of grab-sample particle counting.

3. Running the Sample

Once the sample has been collected, care must be taken to avoid contamination from the grab-sample unit. Most units pull the sample through a small piece of flexible synthetic tubing. Particles from previous samples can collect on the outside and inside of this tubing, skewing the resulting data. The outside of this tubing should be rinsed with particle-free water, or excess sample. The inside should be flushed out by running excess sample through the unit before the actual sample data are collected.

Before running each sample, the grab sampler should be flushed out with 50 to 100 ml of particle-free rinse water. If this is not available, collect an extra beaker of sample from the filtered water source and use it before running settled or raw samples. Before running filtered samples, flush the unit thoroughly with excess sample from the same source.

Order your samples so that the lowest concentration (filtered or finished) samples are tested first, then the next lowest (settled), and then last the highest concentration (raw). This will help minimize problems due to cross-contamination. Establish a baseline with the particle-free or other flush water to determine when contaminants have been minimized. Observe the counts whenever anything is being run through the particle counter, to improve the "feel" for what is going on.

Run several tests on the same sample. The guidelines for the State of California suggest that three tests should be run, with the results for each test within 10% of the average of the three. This is a good rule of thumb. Collect enough sample to run at least five tests of 25 ml or more (do not forget extra sample for flushing the unit). The first test may be higher due to sampler contamination, and the last due to settling of particles. This should leave three tests with counts within a few percent of each other, if all is well. If the results are not consistent, collect and run another sample. Particle counters may not count consistently from unit to unit, but most individual units perform well in terms of repeatability. If consistent results cannot be achieved, the problem will most likely be found in the sample-handling technique. Of course, a consistently poor technique could result in consistent but incorrect results, so nothing is guaranteed.

4. Sample Dilution

Samples requiring dilution are better suited to grab-sample particle counting. However, dilution is bound to introduce a certain amount of error, so it should only be performed when necessary. To understand when dilution becomes necessary, we must review briefly what is referred to as coincidence error.

a. Concentration Limits of the Particle Counter

The concentration limits of the particle counter are usually specified as a percent coincidence error. A typical unit might be rated for 10% coincidence error at 14,000 particles/ml. This means that the actual number of particles counted should be within 10% of the number counted by the particle counter when the concentration of the sample is 14,000 particles/ml. As concentration is increased, this coincidence error increases. Figure 5.1 shows this relationship. Note that the error increases rapidly above a certain point. It is easy to be fooled because the particle counter will only count to a certain level no matter how many particles are passed through it.

As a rule, the coincidence errors specified are lower than what actual experience dictates. The effective concentration limit of a particle counter specified at 14,000 particles/ml may be closer to 8000 or less. One way to spot coincidence problems is to compare the ratio of counts in the smallest size range to those in the next higher range. As coincidence error increases, the smaller particles will be counted simultaneously, resulting in a lower count total in the smallest range. The concentration of particles in water is inversely proportional to size, increasing exponentially as size decreases.

The best way to proceed is to determine the practical concentration limit of the particle counter grab sampler, and then back off that number another 25% or so. This concentration then becomes the cutoff point for diluting the samples. For example, if the effective concentration limit is 8000 particles/ml, then use 6000 as the break-off point for diluting the sample. The following method provides a good way to determine this practical limit, as well as practice for improving dilution skills.

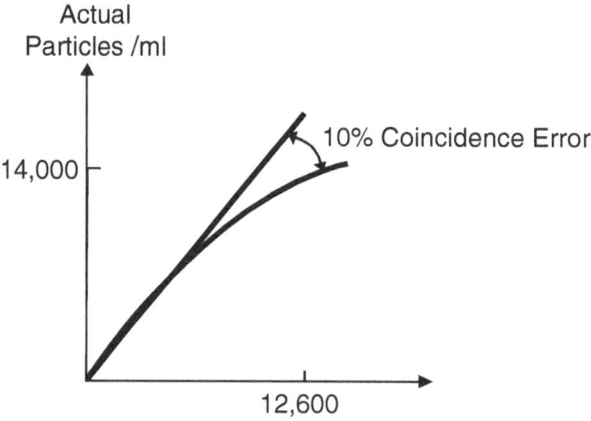

Figure 5.1   Coincidence error.

Table 5.1   Dilution Test

Sample Dilution	Total Counts (all size ranges)
Initial sample at 80% of specified limit	8000 particles/ml
Sample diluted at 1:1	4000 particles/ml
Sample diluted again at 1:1	2000 particles/ml

Notes: Sensor-specified coincidence limit of 10,000 particles/ml. Dilution with particle-free water.

b. Dilution Test

As mentioned above, repeatability is the best measure of proper sample handling, and the same goes for dilution. To determine the practical concentration at which one should dilute, follow these steps:

1. Run a sample with no dilution, or with the minimum dilution necessary to produce a concentration 20% lower than the specified concentration limit of the particle counter. This becomes the baseline.
2. Dilute the baseline sample at a ratio of 1:1. Run it through the particle counter, and record the data. Compare the data with the baseline data, for both counts and size distribution.
3. Perform a second 1:1 dilution, and repeat the test. The measured concentration should now be about 20% of the specified concentration limit of the particle counter. If it is higher than that, then the specified limit is probably inflated. If it is a lot lower, the dilution was probably not done properly. This procedure is reviewed in Table 5.1, with some simplified numbers.
4. Overview. The goal of this procedure is to produce consistent data at various levels of dilution. When the data are corrected to account for the dilution ratio, the measured data should be consistent both in terms of counts and size distribution. Error will occur if the particle counter is overconcentrated, and will also occur if the sample is diluted too much. Once the "reliable" concentration limit of the particle counter is determined, samples should be diluted just enough to keep the concentration at that amount, and not less.

c. Diluents and Background Counts

When a sample is diluted, allowance must be made for the particles present in the diluent. As the dilution ratio is increased, these particles become an increasing part of the measured concentration. The chemical compatibility of the two samples must also be taken into account, as certain chemicals may cause particle coagulation or breakup, thus skewing the data. For raw or settled water samples, the lowest-concentration filter effluent sample can be used, if particle-free water is not available. Finished water has been chlorinated, and will usually have a few more particles than a good filter effluent sample.

D. DATA HANDLING

As touched on above, grab-sampler data handling is a cumbersome task. Any number of approaches are available for organizing grab-sample data, and this makes a standardized approach difficult. Hence, none of the manufacturers has developed useful data-handling software for grab sampling.

The first recommendation is to use computer software to store and display the data. Paper tape printouts are a nuisance, and most units are designed to off-load data directly to a personal computer. A good spreadsheet program will be necessary. Once a method for organizing the data has been developed, a macro can be designed in the spreadsheet to automate the data manipulation.

In most cases, the data should be organized by sample location (filter 1 effluent, filter 2 effluent, settled, raw, etc.) and time of sample collection. In this way, some sort of trend can be developed which will provide a framework for interpreting the data. If samples are taken frequently enough, the data can produce a reasonable picture of a complete filter run. Odd or unexpected increases in counts at a certain point in the filter run may be the result of poor sample handling, but if a pattern emerges over several filter runs, the data may be considered reliable.

Just as in the case of online particle counting, trending the grab-sample data with other plant parameters will add to its value. If these data are available in a usable file format that can be imported into a spreadsheet or database program, the task is made easier. The more data that are available, the more complete the picture will be. It will not be possible to create too much useful data with a grab sampler, only too much to handle efficiently.

In the section on sample handling, we recommended that at least three test runs producing results within 10% of the average of those three were necessary to ensure accurate data. It is best to use the average of these three runs as the data point for each sample. This will greatly simplify data handling.

E. PREPARING A WORKABLE APPROACH

With all the pitfalls and problems related to grab-sample particle counting, a well-thought-out approach is a necessity. A lot of time is required to achieve meaningful results, and several operators will be involved. Any number of methods may be employed to keep this task manageable. The following guidelines are presented to offer an example, and can be adjusted to meet the needs of the particular situation.

1. Operator Training

The first step is proper training of the operators who will be performing the grab sampling. Consistent procedures must be implemented to minimize the problems inherent to particle counter grab sampling. This training should include the following:

a. A good understanding of how particle counters work and how they are used in drinking water treatment.

b. General laboratory sample-handling and testing skills.
c. Particle counter sample collection and handling. This involves a lot of "hands-on" work with the grab sampler. Have each operator collect and run the same sample at the same time to work on consistency.
d. Computer spread sheet or database software manipulation.

2. Procedures

Develop a procedure that involves regular testing of the relevant points in the process stream. This will probably be done only once every 4 hours or so. Make sure that allowance is made for shift changes so that a regular collection pattern is maintained. In most cases, the same operator should collect and run the samples, to ensure that the sample is handled properly. This will allow operators to learn to correct their own mistakes.

3. Data Presentation

Maintain a continuous display of data to allow the operators to work within an intelligent framework. If the new shift operators can see the results obtained by the previous shifts, they will be better prepared for what to expect from their testing. Particle count data means little in isolation, and the operators will be more inclined to perform grab sampling carefully if they are helping to build upon something they can see and understand.

4. Preventing Entropy

Make sure that all sample beakers and other materials are properly cleaned and stored after each use. The next shift should readily find everything in its place.

5. Maintaining a Consistent Sampling Pattern

Collect samples in the same sequence, which will probably be determined by the physical layout of the plant. If a separate beaker is used for each source, label it accordingly.

F. CONCLUSION

It is not likely that a highly rigorous full-plant collection routine will be used for very long, if at all. Too much staff is involved, and if particle counting is found to be valuable enough to warrant the effort, an online system should be installed. Grab sampling will then be used for pilot testing, calibration verification, or some other specialized test program. The approach taken and procedures developed will depend upon the application.

PART II

Understanding the Technology

In this part of the book we turn our attention to a more in-depth look at particle counters, from the basic construction of the instrument to the features of the complete particle counting system. Particle counters are presented generically in this part, while in Part III the actual manufacturer's product offerings are presented. The chapters for Parts II and III are designed to run in parallel, to allow for easy cross-reference.

CHAPTER 6

Specifications

All instruments are designed for a specific task. Specifications are the measure of how well the instrument performs the various parts of that task. Some specifications are common to all types of instruments (dimensions, weight, power requirements, etc.) and are easily understood. Specialized instruments have specifications that are unique to the task for which they are designed. In many cases, common specification terms such as resolution and sensitivity must be understood in relation to the unique properties of the application.

In this section, the primary specifications employed for evaluating particle counter performance are defined and discussed. Understanding of these few specification parameters is essential to making informed decisions about particle counters. If these are not clear at first reading, readers should refer back to this section as they continue through the book. It will become evident that these specification parameters are interrelated, and will only be understood completely when the whole picture of particle counting is clear.

A. SENSITIVITY

Sensitivity is the smallest measurable amount that an instrument can detect. In the case of the particle counter, this refers to the smallest particle size that can be reliably detected. The general rule in the industry is that a particle can be "reliably" detected and measured if it produces a signal with a minimum 2 to 1 signal-to-noise ratio.

B. SIGNAL-TO-NOISE RATIO

All electronic instruments are subject to various forms of "noise," or random, meaningless electrical signals. The term *noise* comes from the analogy to sound levels in a given space. Noise is meaningless sound, and every space where sound can exist will have some level of noise present. Our ability to discern specific sounds depends upon those sounds being at a higher level than the noise present. The amount by which the specific sound exceeds the noise level is called the signal-to-noise ratio.

The sound found between stations on a radio is a good example of electronic noise. Even the highest-fidelity audio system produces some of this type of noise, which becomes audible as the volume is increased. Every type of electronic device is affected by this phenomenon. In the case of the particle counter, noise is produced by the electronic detector circuit, the laser light source, the power supply, and is induced by other devices. The imperfections in the structure of the flow cell windows cause minute scattering and blurring of the laser light source. The combination of all these factors results in a noise level that can be observed and measured with an oscilloscope. (An oscilloscope displays electronic signals in a manner similar to a television.) See Figure 6.1 for a representation of noise as displayed on an oscilloscope.

This noise level is measured with no particles in the water flowing through the particle counter sensor. (This is not exactly true. There are millions of particles well below the detection limit of the particle counter sensor. Some of these contribute to the noise level in a small way.) Once the noise level has been determined, particles of a known size can be sent through the particle counter sensor, and measured. The resulting amplitude produced by the particles is compared with the noise amplitude to determine the signal-to-noise ratio. See Figure 6.2.

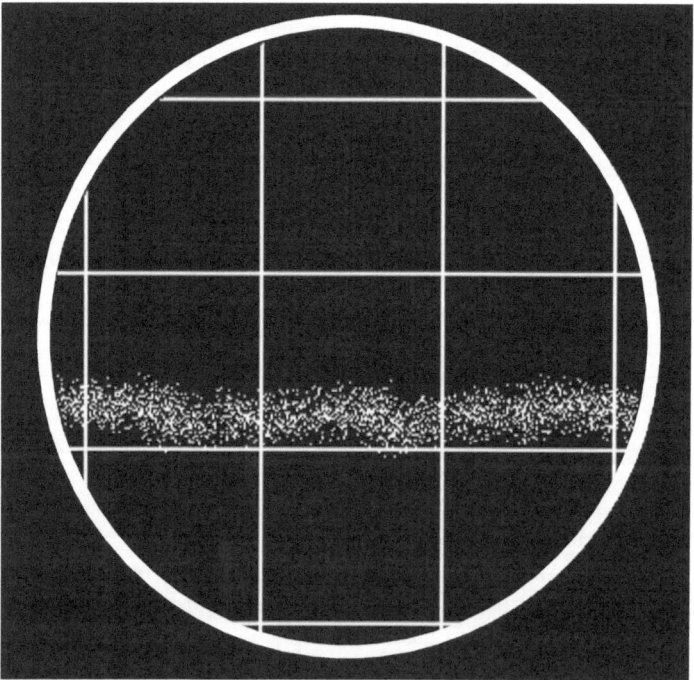

Figure 6.1 Noise on oscilloscope.

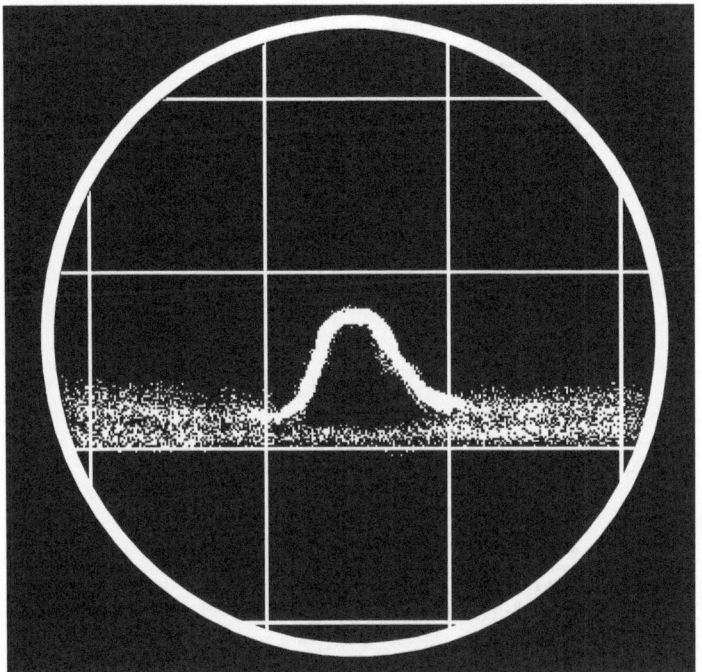

Figure 6.2 Pulse signal-to-noise measurement.

As mentioned is Section A, the smallest particle that can be reliably measured must produce a signal amplitude at least twice that of the noise, or a 2 to 1 signal-to-noise ratio.

C. RESOLUTION

This term refers to the degree to which an instrument can distinguish between differences in the object of measurement. In particle counting, resolution refers to the accuracy with which the sensor can distinguish and measure differences in particle size. This is stated in terms of percentage. A (+ or −) 10% resolution error would mean that particles ranging in size from 4.5 to 5.5 μm could be measured as 5 μm in size.

Resolution measurement is a somewhat sticky subject in the particle counter industry. It will be discussed in further detail in relation to calibration and other issues.

D. COINCIDENCE

Coincidence was discussed in Chapter 5 in relation to sample dilution. Basically, it is the "coincidence" of two or more particles in the sensor view volume at the same time, which is a function of the concentration of particles in the sample. There are other factors that affect this, but they are beyond the scope of this book. Coin-

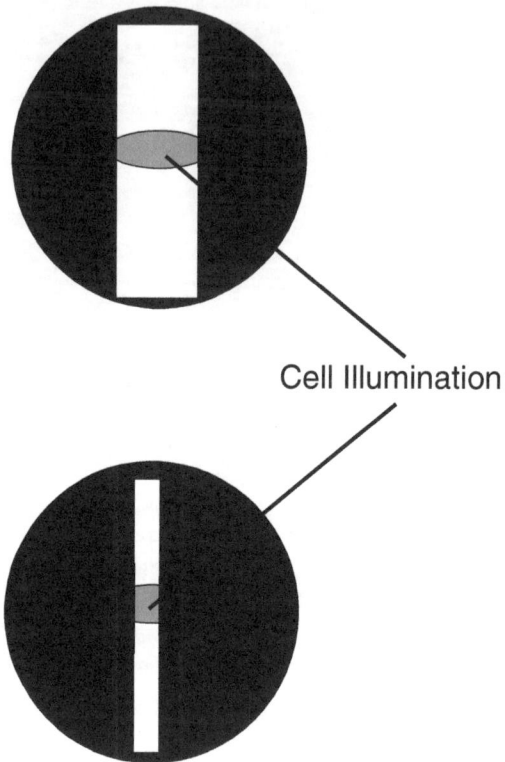

Figure 6.3 Resolution error as a function of cell width.

cidence is usually presented as the number of particles for which the measured counts fall within 10% of the actual number of particles present. This number is the total number of particles per milliliter that are measurable by the sensor.

E. SIZING RANGE

This measure may appear under various names, but it is basically the range of measurable particles that can be sized and counted by the particle counter. It extends from the minimum size particle (usually 2 μm) up to 400 μm or so. The upper limit is generally not too important, as most particles of interest will fall well below 50 μm.

F. SAMPLE FLOW RANGE

All particle counter data must be collected for a known volume of sample, which usually means the flow rate must remain constant or be measured continuously. The specified sample flow range refers to the flow rates at which the particle counter sensor can measure particle size accurately. If the particles are flowing through the sensor at too high or too low a velocity, they will not be sized properly by the detector electronics. The reasons for this are covered in Chapter 7.

G. FLOW CELL DIMENSIONS

The flow cell is the pathway through which the sample must flow. A larger flow cell will clog less often and be easier to maintain. On the other hand, a smaller flow cell will permit a higher concentration sample to be measured with less coincidence error. A trade-off is necessary.

Resolution is also a function of flow cell size. The laser light source will not maintain an even intensity across the entire flow cell, and this error grows with the width of the flow cell. Particles passing through the edges of the beam will not produce the same amplitude pulse as particles of the same size passing through the center. See Figure 6.3.

H. VOLUMETRIC

The term *volumetric* refers to the coverage area of the laser light source. If the beam covers the entire width of the flow cell, the sensor is volumetric. If only a fraction of the width is covered, then it is non-volumetric, which is sometimes called in situ. An in situ arrangement allows for a much higher concentration of particles to be measured because it reduces the effective volume of the flow cell. The drawback is that resolution is diminished because particles may partially pass through the beam and thus be undersized. Figure 6.4 provides an illustration of this.

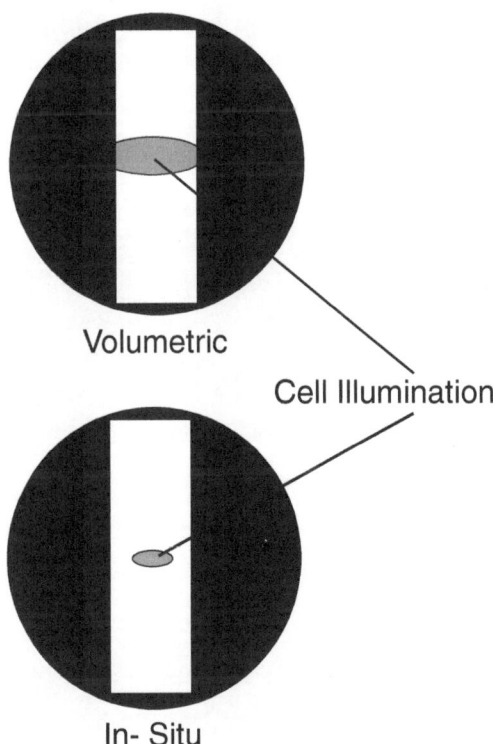

Figure 6.4   Volumetric and in situ sensors.

CHAPTER 7

Particle Sensor Construction

A. FLOW CELL

The flow cell is a foundation piece of the particle counter sensor. It is generally made out of aluminum or stainless steel or even plastic. Stainless steel is more expensive primarily because it is much harder than aluminum and thus more difficult to machine. Aluminum must be coated or specially anodized to prevent pitting when exposed to water. The sample flows up through a channel formed by the flow cell windows and the slit through the flow cell. The sample tube fittings are mounted into the flow cell on both ends of the slit. In most cases, the laser/optics assembly and the detector circuit are bolted to either side of the flow cell. See Figure 7.1.

The flow cell must be rigid enough to hold all these pieces together without allowing any alteration in alignment, which would result in the particle counter sensor losing calibration. The cell windows must fit tightly and be sealed to prevent sample leakage.

The materials of construction are not important, as long as the functional integrity is maintained. The ability to withstand repeated cleanings and clog removal is a necessity.

B. CELL WINDOWS

The cell windows employed are usually round disks of synthetic sapphire with flat, polished surfaces. These windows are seated on either side of flow cell slit and sealed with O-rings to create the flow cell path. Quartz can be used for this type of window as well, or can be drilled and polished to produce a single four-sided piece that contains the flow cell path as well. The laser light beam is sent through the center of the window through the flow cell path and then out the other side, where it strikes the detector.

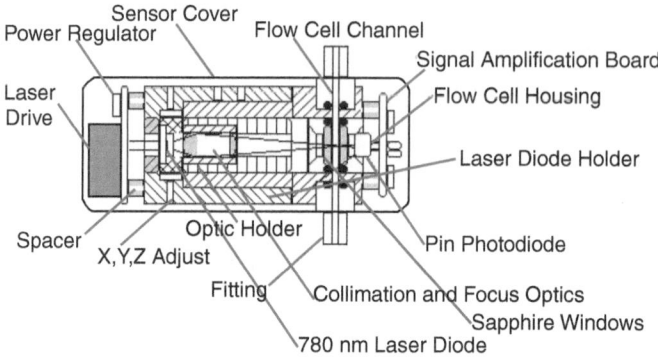

Figure 7.1   Particle sensor. (Courtesy of L & H Environmental, Inc., Roseburg, OR.)

The cell window materials vary in optical quality. Small imperfections in the material diffuse the light and are a source of resolution and sizing error. Quartz may have fewer imperfections as far as optical quality is concerned. It is a softer material than sapphire, and thus can be bored and polished to create the single window piece. This softness accounts for the fact that it can be scratched from cleaning, or from contact with particles over a long time period. Different grades of sapphire are available. The entrance window (the side where the laser enters before passing through the flow cell) must be of a higher grade than the exit window. The beam must be tightly focused where it contacts the particles. Since the detector is measuring total light energy, it is not greatly affected by some diffusion in the exit window.

The entrance window is covered with an antireflective coating to keep the light energy from reflecting back into the laser. Lasers produce light by bouncing light energy back and forth at a high frequency between mirrored surfaces, which amplifies the light intensity to produce the desired output beam. If light is reflected back into the laser, it can disrupt this process, and cause variations in the beam intensity. This is another reason that aluminum flow cells are anodized a dark color, as the shiny metallic surface can increase the problems with reflections. Stainless steel flow cells cannot be coated in this way, and thus require finer alignment to minimize reflections.

C. SAMPLE FITTINGS

A variety of sample fittings are available, and are usually of concern only as a matter of convenience or preference. Ease of accessibility is important. Plastic or nylon fittings should be easily replaceable, as they will break periodically. Some older sensors still in use have pressure fittings that can alter the optical alignment if tightened too snugly.

D. LASER/OPTICAL ASSEMBLY

The laser light source is passed through a series of lenses to produce a thin, even beam across the flow cell. These lenses must be rigidly mounted to maintain the proper alignment. This is usually accomplished by mounting the laser and other optical lenses on a single block or in a tube. This "subassembly" is then bolted to the flow cell.

The detector electronics assembly is mounted to the opposite side of the flow cell. Once the entire assembly is in place, final adjustments are made to produce the desired alignment.

CHAPTER 8

Particle Counter Electronics

The particle counter contains four primary electronics assemblies. The first two are located in the sensor assembly. They are the laser driver circuit and the detector circuit. The other two are usually located in the main enclosure. These are the counting electronics and the power supply. A brief description of each follows.

A. LASER DRIVER

As discussed in Part I, one of the major advantages of the particle counter over the turbidimeter is the stability of the laser light source. The laser diode is a major improvement over the incandescent light bulbs used in early particle counters. It contains no filament to shift or burn out, and can last for decades. The only potential problem for the laser diode is fluctuation in the intensity of the light output. This problem is corrected by using an electronic feedback circuit to maintain a constant beam intensity. Each laser diode contains a built in photodetector that converts light to an electrical signal. This signal is monitored constantly by the laser driver circuit. If the signal begins to drop, more power is applied to the laser diode, and vice versa. In this way, the output of the laser diode is kept constant.

B. DETECTOR CIRCUIT

The laser light source is sent through the flow cell to the detector circuit. This circuit consists of a photo-diode and amplifier circuit. When light strikes the photo-diode, it is converted to an electrical signal and amplified. In the case of the light-blocking sensor, the photodiode is constantly illuminated, and particles passing through the light beam cause a temporary dip in the intensity of the light striking the photodiode. The detector circuit inverts the output of the photodiode so that no voltage is output when no particles are present (except for electronic "noise"). When a particle of sufficient size passes through the light beam, the output of the detector circuit rises to the equivalent voltage level.

The voltage value that corresponds to a given particle size is dependent on the characteristics of each individual sensor. There is no "correct" output voltage. Enough amplification is supplied to achieve a sufficient signal level for the smallest size particle (usually 2 μm). Once adjustments have been made and the 2 to 1 signal-to-noise ratio has been achieved, the voltage values corresponding to each particle size are determined by measuring the output when particles are sent through the beam.

The output signal is amplified only enough to be sufficient for the counting electronics. Increasing the amplification does not improve the signal-to-noise ratio beyond a certain limit. It is the same thing as turning up the volume on a weak radio signal. The noise is amplified as much as the signal.

Chapter 1 discussed the flow rate range of the sample. The flow rate must be kept within a certain range for the particle counter to size the particles properly. Obviously, at the low end, the sample must move at a sufficient velocity for the particles to get through the flow cell. At the upper end, the particles can move too quickly through the beam to be properly sized. This is due to a phenomenon known as bandwidth limitation. Without getting deeply into electronics theory, a simple analogy will serve as an illustration.

Even the fastest automobile cannot go from 0 to 60 mph in zero time. Gravity, friction, wind resistance, and a host of other factors limit its acceleration. Electronic signals are analogous. While they are much, much faster than any car, they cannot change voltage levels instantaneously. As the particle passes through the light beam, there is a speed at which it can pass completely through the beam before the output of the detector electronics can reach the full value that normally corresponds to particles of that size. For this reason, particle counters are calibrated at the same flow rate for which they are designed to operate.

(For those interested in the electronic reasons, the bandwidth limitations are primarily due to the capacitance of the photodetector. The impedance of a capacitor varies with the frequency content of the electronic signal applied to it, resulting in a loss of amplitude at higher frequencies.)

A wide flow range is not too important in itself, but could be viewed as evidence of superior detector circuit design. Ideally, the bandwidth of the detector circuit should be high enough so that the amplitude of the pulses is not being reduced at the standard flow rate.

C. COUNTING ELECTRONICS

Several types of counting electronics are found in the particle counter industry. All of them serve the same end, which is to sort and count the pulses produced by the particle counter sensor. They must collect these values over a given time, sort and totalize them, and then communicate the resulting data to an external device such as a computer, chart recorder, or some other type of monitoring device. We will begin a presentation of the types of counters available with the simplest, and move up in complexity from there. The output of the counter electronics will either be in the form of a 4 to 20 mA signal or a digital output designed for computer interface. As each of these is discussed in detail elsewhere in the book, the present discussion will focus on the counting function of these circuits.

Particles passing through the particle counter sensor produce a voltage pulse output that varies in amplitude in proportion to the size of the particle. The counting electronics then sorts these pulses and counts them according to size. To achieve this, the counter must be able to discriminate between pulses of different amplitude. Two methods are employed. The simplest utilizes what is called a voltage comparator input stage. The more complex approach involves analog-to-digital (A/D) conversion.

1. Voltage Comparator

A comparator is simply a small circuit that "compares" an input signal to a fixed reference voltage. If the input signal is greater than the reference signal, the output of the comparator goes "high" or "ON." If it is lower than the reference, the output remains "low" or "OFF." The output will only remain high as long as the input signal remains higher than the reference level. When a voltage pulse from the particle counter sensor is applied to the comparator input, the output of the comparator will go high for the duration that the pulse amplitude exceeds the input reference. In this way, a series of particle pulses can be translated into a string of high outputs which are then totalized by a counting circuit. Several comparators are combined to provide counts for different sized particles.

For example, consider a particle counter with size ranges set for 2 to 5 μm, 5 to 15 μm, and 15 μm and larger. This would require three voltage comparators. These are pictured in Figure 8.1. Imagine that these comparators are arranged like steps on a ladder. The lowest step would be set to match the voltage equivalent to a 2-μm particle. (This value is determined by the calibration method in Chapter 14.) The second step would be set for 5 μm, and the third for 15 μm. Ignoring the "electronic noise" of the sensor output, we will assume that when no particles are present we have zero volts output. (This is referred to as "ground" potential in electronics, and we can think of it as the "ground" upon which our "ladder" is resting.)

When a particle passes through the sensor, an output pulse is produced that corresponds to the size of the particle. If this pulse is "higher" in amplitude than the first step of our ladder (the 2-μm threshold), the output of the first comparator will go high. If the pulse amplitude exceeds the value for 5 μm as well, both the 2 and 5 μm comparators will turn ON or go high. As each particle passes through the sensor, a unique pulse is produced, which will trigger one or more of these comparators if it is greater than 2 μm. The counting circuit then totals the number of times each comparator has been turned ON or "triggered." If two comparators are simultaneously triggered, the counter will place a count in the higher size range.

These counts are totalized for a given time period, which corresponds to a fixed sample volume. Typically, the counts might be collected for 15 seconds (a 25-ml volume at a flow rate of 100 ml/minute). The counter will then ignore the particles

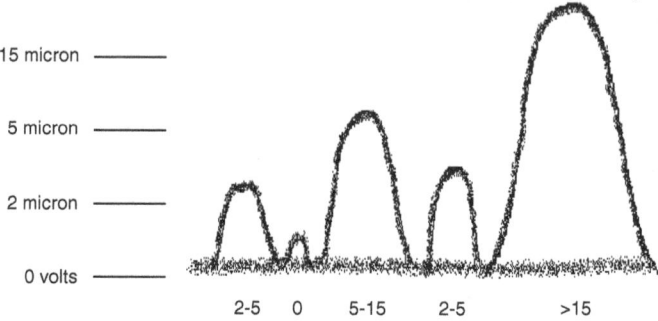

15 micron

5 micron

2 micron

0 volts

2-5   0   5-15   2-5   >15

Figure 8.1 Comparator counting method. Particle pulses counted and sorted by amplitude. Picture represents sequential pulses in time

for several seconds to allow time to sort and communicate the data. Data are typically communicated every minute.

The comparator provides a simple way to convert the particle size information from an analog pulse to a digital count value. It is a very basic form of analog-to-digital conversion. The "resolution" of the counter depends upon where the size thresholds are set. In most cases, only a few size ranges are provided when comparator inputs are used.

The accuracy of the counting electronics will be determined by how precisely the reference signals are set for each particle size threshold. The voltage comparator has a certain degree of imprecision, meaning that it will trigger on voltages within a few percent of the exact voltage for which it is set. It should also be obvious how "coincidence" error can affect the counting accuracy. If two or more particles pass through the sensor light beam at the same time, only one pulse is produced, and it will correspond in amplitude to the aggregate size of the particles. The comparator will be triggered only once, and if the measured size is large enough, the wrong threshold could be exceeded, causing a sizing error as well. If the particle concentration is too high, a second particle could pass into the beam before the first is completely out of it, and the output pulse may not drop low enough to turn the comparator OFF. In this case, only one particle will be counted instead of two.

It should be noted that a minimum 2 to 1 signal-to-noise ratio is necessary for preventing extraneous counts in the lowest size range. If the "noise" level becomes too close to the lowest comparator threshold, the comparator will respond to the spurious noise signals, making the count data unreliable. Comparators can trigger on pulses less than a millionth of a second in duration, so they are easily set off by electronic noise.

The particle pulses themselves have a certain amount of noise "riding" on them, which causes fluctuations in the pulse amplitude. This can cause the comparator to trigger multiple times on a single pulse. This problem is reduced by implementing what is known as hysterisis. This is simply turning the comparator off at a lower pulse amplitude than what is required to turn it on. If the "gap" between the ON and OFF levels, or hysterisis, is larger than the noise level, the false triggering caused by noise can be greatly reduced or eliminated. However, since the turn-off level is lower than the turn-on level, the signal-to-noise ratio is once again a critical factor.

2. Setting Comparator Size Thresholds

Several options are available for setting the size thresholds for comparator counting circuits. The simplest type are factory preset, and are not designed for field adjustment. This means that any changes in the size ranges measured by the particle counter will have to be set at the factory, or by factory-trained field technicians. In most cases, these settings cannot be verified in the field. They have to be checked and adjusted during calibration to ensure proper operation.

User-settable thresholds are also available. The simplest of these provide a potentiometer setting and voltmeter to allow the operator to change and verify settings. More complex systems provide digital-to-analog converter outputs to set the comparator thresholds. These require a computer interface where the desired size ranges are input by the operator, and internally adjusted by the particle counter electronics. The main advantage of user-settable thresholds is that changes can be made on site if new regulations necessitate. In all cases, the accuracy of the threshold settings and counting electronics should be verified during calibration.

3. Analog-to-Digital Conversion

The more advanced counting circuits employ analog-to-digital (A/D) conversion of pulse amplitudes into count data. A/D conversion provides better sizing resolution than the simple comparator method. In the example of a comparator input circuit presented in Figure 8.1, particles could only be "resolved" into three different ranges. With an A/D converter, the potential exists to resolve data into thousands of size ranges. (See Chapter 4 on 4 to 20 mA signals for a discussion of resolution.) While this is not practical for most applications, there are areas where this is important, such as in calibration. As the development of high-speed electronics progresses, new applications may be discovered for utilizing the full capability of high-resolution particle sizing.

The A/D circuit is used to convert the peak amplitude of each pulse into a digital value corresponding directly to it. These values are collected and processed to provide the counts in each of the selected size ranges. Just like the simpler comparator-type counters described above, pulses are collected and stored for a fixed time period. The counter is then shut down and the counts are processed and communicated.

This type of counting circuitry requires a much faster and more complex processor than the basic comparator circuit. In all cases, the thresholds must be set via a computer interface, as no manual trimpots or adjustments are employed. The incoming particle pulse is captured by a "peak detector" circuit that follows the amplitude of the pulse to its highest point and stores that value long enough for the A/D conversion to take place. The A/D converter consists of thousands of comparators, which can resolve the pulse amplitude to within less than a millivolt in some cases.

The A/D counter circuit is still susceptible to the errors described for the simple comparator circuit. Unlike those circuits, the A/D counter is less susceptible to spurious noise. Several milliseconds are required for the A/D to resolve an individual pulse, and noise spikes will not normally last that long. In most cases, the A/D counter will not count as many particles as the comparator circuit as concentrations increase. One big advantage is that the number of size ranges is not physically limited in the A/D counter, as it is for the comparator circuit. This could become important if new discoveries increase the need to monitor several size ranges.

For most applications, the type of counter circuit will not be an issue of great importance. As far as the operator is concerned, the units operate the same way from a functional standpoint. As standards are developed and become defined to a greater degree, these differences may become more important. The high-resolution A/D counter circuit provides the potential for autocalibration and count matching, if these issues ever become a priority in the industry.

4. Pulse Height Analysis

A special type of A/D converter counter is known as a pulse height analyzer (PHA). This type of counter is used primarily for particle counter sensor calibration, but can be used for other applications in drinking water treatment. The PHA is used to provide a histogram of all the particles in a given volume of water. A "histogram" is a sort of graphical history of the particles present in that volume of water. The PHA is used to record the counts in each of several thousand size channels. These counts are displayed on an x–y graph as shown in Figure 8.2.

For calibration purposes, the counts are displayed vs. the actual voltage amplitudes of the particle pulses. The data can also be displayed as a function of the number of counts vs. particle size. In most cases, the PHA is used in conjunction with a grab sampler or calibration station, as too much data would be produced by a continuous, online particle counter.

The PHA could be used to develop an extensive picture of particle size distribution in a settled or raw water source. It could also be used to determine sensor coincidence and proper dilution ratios as discussed in Chapter 5.

D. POWER SUPPLY

The power supply is usually a standard, off-the-shelf unit which is not designed by the particle counter manufacturer. Two types of supplies may be used. The most popular is the switching supply. Some units may incorporate a linear supply. A linear supply uses a stepdown transformer to achieve the desired voltage output, while a switching supply utilizes capacitors and transistor switches. Linear supplies can be less noisy, but are heavier and less efficient (meaning they draw more power and run hotter). Switching supplies also have the advantage of accepting a wider range of input voltages. This reduces problems due to power "brownouts" or fluctuations common in large treatment plants.

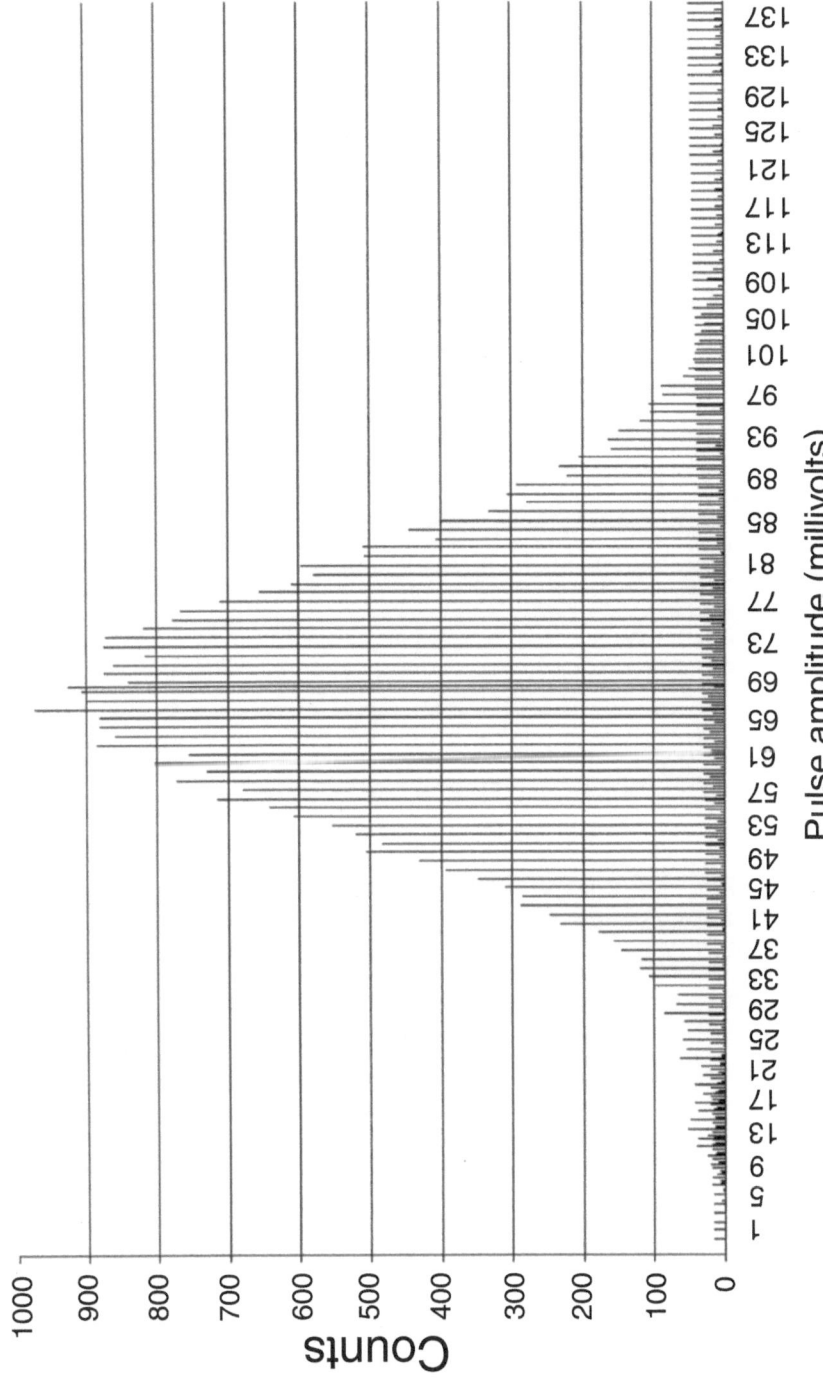

Figure 8.2 Pulse height analyzer data. (Courtesy of Pacific Scientific Instruments, Grants Pass, OR.)

CHAPTER 9

Auxiliary Features

Most particle counters provide diagnostic signals and alarms. In many cases, particle counters are used in conjunction with other instrumentation, and most can accept signals from these instruments to be trended along with the particle counter data. Flow control is also critical, and several methods are provided to achieve accurate flow control. In some cases, flow alarms can be signaled by the particle counter via the computer data collection system.

A. DIAGNOSTIC SIGNALS, ALARMS, AND DISPLAYS

The most common alarm found on particle counters is the cell condition alarm. This is used to alert the operator that the particle counter sensor has been fouled or clogged. This signal can be returned as a "good/bad" signal, or as a numerical output. The later is preferable, because it provides advance warning of possible problems. This signal is standard on digital output particle counters, and should be provided for with 4 to 20 mA output units as well. This alarm should be displayed in the central control room or on the data collection system display, as it requires immediate response.

Count alarms are usually programmed in to the data collection system, and are not part of the particle counter hardware. These can be useful as operational tools once a good deal of operating experience has been gained. Until that time, they should be turned off. Since no "ideal" count number has been established, these alarms are confusing to inexperienced operators. Alarms usually signal that something is wrong, and should never be set to arbitrary values. One exception is 4 to 20 mA output units. Since the output signal can only rise to a certain value, the operator should be warned when the counts exceed the output range of the current loop. High counts may indicate a filter breakthrough or process upset, and should never be allowed to "max out" on the 4 to 20 mA output scale without warning.

If a 4 to 20 mA signal drops below 4 mA, there is obviously a problem with the particle counter or cabling. The receiving equipment should be designed to alarm when this occurs, as the particle counter may be dead and unable to provide an alarm. (In most cases, the cell condition would drop to zero as well, indicating a big problem.) Digital output units arc "polled" by the data collection system. If they quit responding, the data collection system will usually inform the operator with an alarm message.

Some particle counters provide local display of data and alarms. These are primarily of use during installation and maintenance. The particle counters are usually mounted down in the filter gallery or in other out of the way places, so there is little need for local display for day-to-day operation. The ability to view the cell condition or flow rate is useful when cleaning the particle counter and setting up or verifying a flowmeter. As mentioned in Part I, individual count values are of little practical value outside of the overall trend.

B. SAMPLE FLOW REGULATION

Sample flow monitoring was discussed in great detail in Part I. Many particle counters provide inputs for flowmeters or low-flow alarms. Some even provide flowmeters as standard or optional equipment. Flowmeters usually provide an analog output signal, which can be connected to an auxiliary input on the particle counter counting electronics. A low-flow alarm circuit may be part of the counter electronics, or may be a self-contained unit that requires an input on the particle counter.

Some particle counters may use the flowmeter signal to determine the number of counts per milliliter directly. Others rely on the data collection system for this calculation. In either case, high- and low-flow alarms should be provided to prevent the flow from exceeding the acceptable limits of the particle counter.

C. ANALOG INPUTS

One of the most useful features of the more-advanced models of particle counters are the auxiliary analog inputs. These provide a means for trending the outputs from other instruments along with the particle counter data. These signals are of primary importance when the particle counter system is a turnkey system provided by the manufacturer. SCADA systems already provide for these inputs, and 4 to 20 mA particle counters would have no use for them.

The number of analog inputs provided will vary from unit to unit. Certainly turbidity and loss-of-head are useful inputs and others may be of importance depending on the process. The most convenient arrangement is to have the analog inputs located on the individual particle counters so that inputs from instruments mounted on the same filter bed can be easily connected into the system. Some systems have separate analog input racks, which may require long cable runs. Several parameters should be investigated when evaluating analog inputs on a particle counter system:

1. As outlined above, the number and location of the analog inputs should be sufficient for all required instruments, and convenient for installation. A spare input can come in handy for future additions, or if an input becomes inoperable.
2. *Resolution*: Make sure the signal resolution meets or exceeds that of the incoming signal; 12-bit resolution will be the highest available, with 10 bit and even 8 bit possible.
3. *Isolation*: Proper signal isolation will prevent damage to the particle counter from improper wiring or signal loop problems. Whenever instruments are electrically connected, there is potential for problems, especially in older plants where the existing wiring is not documented properly. Remember Murphy's law.
4. *Installation Requirements*: Make sure that sufficient space is provided to install the signal wiring properly in the particle counter enclosure. If the particle counter is removed from service for calibration or repair, all auxiliary input wiring will need to be removed and reinstalled.
5. *Signal Termination*: Make sure the proper termination resistor is used.
6. Make sure the loop has sufficient voltage to drive the input.

D. DISCRETE INPUTS

Discrete inputs provide a similar value, and are used in the same way as analog inputs, primarily for turnkey type systems. One of the most useful is a backwash valve position indicator. This is tied to a dry contact on the filter backwash valve, and can signal the particle counter system when the filter is placed into backwash. This provides a way for the data to be properly labeled without attention from the operator. Backwashing will usually increase the particle counts dramatically, and the data generated during this time must be separated from the filtered water data.

The rules for discrete inputs are similar to those for analog inputs. Dry contacts should be used unless otherwise specified. "Dry" indicates that the contacts are isolated from any power sources, and can receive a signal from the monitoring instrument without causing problems.

Some instruments may not provide separate discrete inputs, but rather use an analog input to accomplish this. In such cases, a resistor pull-up circuit is employed to provide two distinct voltage levels. A relay contact is used to switch the input between two discrete levels.

E. ANALOG OUTPUTS

A strong case against using 4 to 20 mA current loops was made in Chapter 4. There are cases where no other option is available, or the system is small enough to keep things manageable. This section provides a brief review of 4 to 20 mA current loop signals, along with some specifics related to particle counting.

1. 4 to 20 mA Basics

Most drinking water plant operators are familiar with 4 to 20 mA signals. Their primary benefit is that currents can carry signals accurately over long cable lengths without loss, unlike voltages, which drop because of the resistance of the signal line. These signals are carried over a "shielded twisted-pair" cable. The cable is shielded and twisted to prevent it from picking up induced noise from electrical equipment in the plant. Most of the receiving devices connected to a 4 to 20 mA current loop have voltage inputs. According to Ohm's law, voltage is equal to current times resistance. All that is required to convert the 4 to 20 mA loop to a voltage is a small resistor. In most cases, the input voltage range of a chart recorder or SCADA analog input is 1 to 5 V. A 250Ω resistor is required to convert a 4 to 20 mA signal into a 1 to 5 V signal.

Another benefit of the 4 to 20 mA current loop signal is that several receiving instruments can be connected to the same signal loop. Since the identical current flows through each device, all can receive the same signal. There is a limit to the number of receivers that can be connected to the current loop. This limit is determined by the voltage used to power the loop. All the voltage drops in the loop must add up to the voltage output of the power supply driving the loop. For example, a loop power supply that operates at 12 V could drive two 1 to 5 V inputs. A 24-V supply could drive four (5 + 5 + 5 + 5 = 20). Allowance should be made for the voltage drop due to the small resistance in the signal wire. These rules should be observed whether wiring a particle counter, analog output, or analog input into a current loop.

2. Signal Power and Isolation

The 4 to 20 mA current loops may be powered directly from particle counter, or a separate supply may be used to power the loop. There are two goals in maintaining proper signal isolation. The first is to protect the particle counter and receiving instrument(s) from damage. The second is to prevent signal interference between current loops. Problems can occur when different current loops are tied to the same ground reference at the transmitting instrument, and connected to different receiving instruments. Ground loops may result in offsets, which create erroneous signals.

3. Output Scaling

Particle counters provide various means of scaling the count output to conform to the limitations of the 4 to 20 mA signal. Some of the units provide a selectable sample time period. The counter counts continuously during the sample period, which might be 6 s, or 60 s, depending on the selection chosen. The internal counter in these units can count only up to a certain level before it reaches its limit. For example, a 16-bit counter can count up to around 64,000 counts. If the sample period is 60 s, and the flow rate is 100 ml/minute, this means that the upper limit of the counter is only 640 particles/ml (64,000 particles divided by 100 ml = 640 counts/ml). This is well below the coincidence limit specified for most particle counters. If the sample period is reduced to 6 s, then the counter can reach a level of 6400 particles/ml. If this type of count scaling is employed, the receiver must be scaled to match the particle data by providing the correct divisor.

A second method maintains a constant sample time, but varies the number of particles counted. For instance, on low-concentration filtered waters, every particle may be counted. As concentration increases, every second, or fourth, particle may be counted. These ranges are switch selectable, and the receiving instruments must also be scaled to match the count range.

Whatever method is employed, both the particle counter and the receiver must be scaled correctly. There is no way for either unit to determine if the other is set properly. In most cases, the data resulting from miss-matched scaling may look "normal." If the counter goes over its limit, an alarm should be provided to warn the operator.

F. DISCRETE OUTPUTS

Discrete outputs are usually provided to drive an external alarm. This type of signal would be more prevalent on simpler 4 to 20 mA output units since the "turnkey" type systems all provide built in alarms on the data collection computer. Discrete outputs should normally be in the form of dry contacts, to provide signal isolation as described in the previous section.

G. ENCLOSURE

The particle counter sensor may be mounted in a separate enclosure from the power supply and counter electronics. The intent is to protect the electronics and optics from the external environment. Since water is passed through the particle sensor, steps must be taken to prevent that water from leaking into the electronics and power supply, which can destroy the equipment and create a safety hazard. Remember Murphy's law: If something can go wrong, it will. If a sample is run into a sealed enclosure, at some point that enclosure will fill up with water.

In most cases, standard NEMA 4 or 4X enclosures should be used. Particle counters should not be mounted outside unless the environment is mild enough to prevent freezing of sample lines, and they are shielded from direct sunlight, rain, and high temperatures.

If external signals or communications cabling are to be run into the enclosure, sufficient space should be provided. The ease with which these cables can be installed and removed becomes important when the instrument must be removed for repair or calibration.

CHAPTER **10**

Serial Data Output

The majority of particle counters provide serial data, or digital, output. Serial data output provides for a much greater amount of information to be sent through a single output. Many units can share a single communication line. This type of data is collected and "decoded" with a computer.

A. BASICS OF SERIAL COMMUNICATIONS

"Serial" communications refers to the fact that data are transmitted in sequential order, one "bit" at a time. Each bit is represented by a voltage level which represents a "high" or "low" logic state. ("Bits" are arranged into "bytes" that are arranged in "words." This computer jargon can be a bit confusing, and is not essential to grasping the ideas presented here.) Basically, serial means "in sequential order" or "one at a time." In a typical particle counting system, several particle counters are connected together on a single twisted shielded-pair cable. This cable is then connected to a computer or other data collection system.

The data collection system (DCS) will collect data from each of the particle counters in sequence. The data are sent out from the particle counter as a stream of bits, which is organized or "encoded" according to a specific pattern, or "protocol." The DCS can decode the data from this stream of bits, and translate the data into particle counts, alarms, etc. The key to the whole process is the speed at which serial data can be transmitted. The data from 30 or more particle counters can be transmitted sequentially in a matter of a few seconds.

Each particle counter is given a unique "address," which allows the DCS to identify and communicate with it. A particle counter is "polled" by the DCS at regular intervals, usually once every count cycle. The data from each particle counter are then transferred to the DCS to be displayed and archived.

Since a lot of data are being transmitted at a high rate of speed, it is inevitable that some data will be lost. Data loss can be the result of interference from induced signals, or "collisions," which can result if two or more particle counters attempt to transmit data at the same time. If the correct data are not received by the DCS, it can request the particle counter to send them again. The DCS is able to determine if the data have been transmitted properly by performing a "checksum" calculation. This checksum is calculated from the data by the particle counter according to a specific algorithm, and attached at the end of the data stream. The DCS recalculates the checksum, and compares it to the one received. If they do not match, the data are requested again.

If no response is received from a particle counter, the DCS will try several more times to communicate with it. If it does not respond after repeated attempts, an error message will signal the operator to correct the problem. As mentioned in Part I, it should be clear that serial data communications provides a more accurate and "robust" way to transmit particle count data. A great deal of data can be transmitted with no loss.

The output signals usually correspond to industry standard RS-485 or RS-232 requirements for serial communications. RS-232 is used for single instruments and short cable distances from the computer or DCS. RS-485 is designed for cable runs of up to 4000 feet, and can support multiple instruments on a single cable. It is important to remember that these standards only refer to the signal voltage levels, and have nothing whatsoever to do with the data protocol. All data transmission involves electrical signal levels of some sort. Never assume that two RS-485 devices will communicate together. This would be akin to assuming that a lawn mower engine will work in a Cadillac, because both run on gasoline.

The data may be transmitted in any type of format, referred to as a "protocol." These protocols are not standardized, and a lot of effort can be expended in "translating" the data into usable form. Many laypeople are confused by the distinction between the industry standard signal formats (RS-485 and RS-232) and the non-standard protocols. To make an analogy, the signal format could be compared to "speaking out loud" and the protocol to the "language" being spoken. Any number of languages can be "spoken out loud." The computer receiving this data must speak the same "language," or have a "translator" (usually referred to as a "driver") that will translate the language.

B. DEFINITIONS

Several new terms were introduced in the preceding section, and will be reviewed here to provide a quick reference. Some are given as working definitions for laypeople, to simplify the concepts presented.

Address: A unique value assigned to each instrument in a system. The address allows the computer or DCS to communicate to each unit individually.

Checksum: A value calculated from the data transmitted, and used by the receiving instrument to verify the accuracy of the transmission.

Driver: Software program used to translate data from one protocol to another.

Polling: The method of communication with several units connected to a central computer or DCS. Each unit is contacted in sequence. It is analogous to the familiar "opinion poll" where information is collected from several people and compiled together.

Protocol: The unique arrangement of data bits that is employed by all the instruments in a system. Corresponds to a "language" in human terms.

RS-485: An industry standard for signal voltage levels used in serial communications. Designed for long cable runs and to permit several units to be connected to a single cable. (RS-232 is for single units having cable lengths of less than 50 ft.)

Serial communications: Transmission of digital data in a sequential format. Several units can communicate over a single cable.

C. SCADA INTERFACE

SCADA (Supervisory Control and Automated Data Acquisition) systems are usually composed of a specialized software package and one or more computers. The software is designed to allow a user or systems integrator to customize the way in which data are collected and displayed to suit the requirements of the particular application. Since SCADA systems are normally used to collect data from most or all of the plant instrumentation, a number of instrument protocols must be supported. Most of the signals will be in the form of 4 to 20 mA current loops. Serial interfaces are becoming more common, with the growing complexity of newer instruments and the need for greater accuracy.

Most of the major SCADA software packages are equipped to handle several different instrument protocols. Many different driver packages have been developed to accommodate popular protocols. The most commonly available are those used for programmable logic controllers (PLCs) or other distributed input/output equipment. Among these are Modbus, Optomux, and Allen Bradley. Some instruments have been designed to emulate these common protocols, to take advantage of their availability. Other protocols have been developed for specific projects, then made available through the SCADA vendors or third-party suppliers.

In the case of particle counters, there may not be readily available drivers. Some of the particle counter manufacturers have developed their own drivers for specific SCADA packages. In most cases, a driver will have to be developed from scratch, or an existing driver modified. Some of the SCADA software suppliers will provide custom drivers, as will third-party developers. The many options and approaches for developing drivers extend beyond the scope of this book. For the purposes of aiding in this process, we provide an overview of how a typical particle counter organizes data for communication. Part III reviews the methods used by the individual manufacturers. In all cases, it is highly recommended that the SCADA supplier, particle counter manufacturer, and system integrator all be consulted before any final specifications or bids are prepared.

D. PARTICLE COUNTER COMMUNICATION PROTOCOL

In addition to the large amount of data produced for each sample, there are timing and control issues involved. Some particle counters run without external control, and will provide the current data when polled by the computer or DCS. Others require more "hand-holding"; i.e., they must receive start and stop counting commands from a central controller. In all cases, the count data are updated at regular intervals, leaving a brief time window for retrieval. The data transfer rate is important, since several units must be polled during each sample interval.

Most particle counters allow for some degree of remote programmability. The DCS can be used to send out size range and sample period settings. This can add another degree of complexity to the driver and SCADA software configuration. In some cases, it may be more practical to use the manufacturer's utility program to set these parameters. Such settings are rarely changed in normal operation. This will usually require shutting down the SCADA system and restarting it after these parameters have been changed.

1. Data Configuration

Data output from the particle counter will consist of several "strings" of information. There are usually four or more particle size ranges, each a separate data point. Most particle counters accommodate analog and discrete inputs, which will create yet another string of data. There are status and alarm points, such as cell condition, low flow, etc. These must be included in the driver interface, as they are critical to operation.

Some particle counters perform the flow rate calculations internally, using either a preprogrammed value or the output from a flowmeter. In these cases, the transmitted data will be normalized to particles per milliliter. Otherwise, the DCS must perform this calculation, increasing the "overhead" that the computer must handle. Scaling or multiplier values may be necessary, depending on the way the data is configured. A 16-bit "word" can be used to transmit integer values up to around 64,000. If the counts are normalized by the particle counter, there must be a way to transmit fractional values.

In most cases, analog input values are returned as voltage levels to be scaled by the DCS. There are too many possible variations in scale ranges to make an on-board calculation practical. Discrete values can be returned in binary or hexadecimal form. A 4-bit word contains 16 binary states, or 64 hexadecimal states.

2. Timing and Control

Particle counters designed to run without any external control will collect sample data at a specified time interval. They are programmed to count for a fixed time, and to remain idle for the remainder of the cycle. If polled while counting, the particle counter will respond with a "busy" signal. During the idle phase of its cycle, it will respond that data are available, and will send out those data when requested.

Particle counters requiring external timing control are initiated with a "universal" start command. A "universal" command is answered by all the particle counters, regardless of their address. This makes it possible for the DCS to start and stop all the particle counters at the same time. The idle period must be sufficient to enable the DCS to poll each particle counter individually to download its count data. In such systems, the flow calculation must be performed by the DCS and not the particle counters. This type system places a greater burden on the accuracy of the driver interface both in terms of timing and overhead.

Some SCADA systems are designed to poll continuously for data. This makes sense when monitoring analog inputs that are continuously varying signals. However, constant polling of the particle counters will cause problems. Any time that a serial instrument is polled, it must respond to the DCS, if only to respond that it is "busy." If polled during a count cycle, the particle counter must briefly discontinue its counting to answer the "interrupt." This is not a problem if it happens only a few times during the cycle. If it happens continuously, a substantial amount of data could be lost. There is no reason to poll the particle counter continuously since the data are only updated every minute or so.

3. Remote Programming

Most of the particle counters currently in production allow remote programming of particle size ranges and other parameters, such as sample flow rate or sample time period. To change these parameters, a string of commands must be sent from the DCS to each particle counter, either individually or as a universal command. In most cases, each particle counter is set to the same size ranges since log removal calculations are only valid for identical size ranges. Flow rates and sample periods are normally the same as well, although in some cases a lower flow rate may be necessary due to insufficient head pressure, or high concentrations of particles.

The size ranges are rarely changed after an initial "trial" period in a new installation. Adding the necessary commands for remote programming to the driver and SCADA software configuration may not be cost-effective. In most a cases, a simple utility program may be used to perform these functions. One of the benefits of remote programmability is that all the parameters of the particle counter can be initialized and verified. Most programmable features are stored in battery-backed-up memory, and are lost if the battery dies. Severe electrical storms or power outages may scramble this memory.

E. COMMUNICATIONS DRIVERS

The communications driver is the link between the SCADA software and the particle counters. It consists of special software code which runs along with the SCADA software. Control commands are issued from the SCADA software to the driver, which translates them into the particle counter protocol. Data returned from the particle counters are translated by the driver into a format compatible with the SCADA software.

The driver and SCADA software must be capable of providing the required timing, control, and data transfer at a fast enough rate to perform accurately. This must be accomplished without inhibiting the rest of the tasks required of the SCADA system. The ever-increasing speed of available computer hardware will make this easier to achieve.

F. SORTING OUT THE OPTIONS

As mentioned in Part I, there are several ways to interface particle counters to SCADA without using a direct driver interface. Most involve some sort of data-sharing arrangement between the manufacturer's turnkey system and the SCADA system. Each of these approaches must be evaluated for the specific application. A brief description of a few of them is provided below:

1. Dynamic Data Exchange (DDE)

Dynamic Data Exchange is what the name implies, a constant "exchange" of data between the particle counter software package and a SCADA software package. The particle counter data are output to a temporary location in the computer's memory (often the Windows "clipboard"). The SCADA software collects the data out of this memory location, and uses the data to update the particle counter data for the system.

DDE is somewhat less complex than a driver, requiring only that the data be presented in a format recognizable to the SCADA software. It can also be less reliable, as two separate software programs are running on the same machine, and sharing memory space. If one "crashes," or the computer memory is corrupted in some way, the data will be lost. The programs are not synchronized, meaning that each has to rely on the other to keep the data moving at the right pace.

OLE (Object Linking and Embedding) has begun to supplant DDE. OLE provides a more robust environment, as the data ("object") can carry "embedded" information about its structure to allow supporting programs or routines to be "linked" together. DDE requires that both programs be configured for the exact structure that the data will take, and the data itself provides no information to support this synchronization. The latest form of OLE provided by Microsoft is known as "Active X."

Recent efforts at improving the reliability and compatibility of data transfer have resulted in the development of OPC (OLE for Process Control). OPC is an industry standard put together by leading providers of instrumentation and control systems in collaboration with Microsoft. It defines the interfaces, methodologies, and require-ments for data sharing between different devices and systems.

2. Networked File Sharing

Some of the turnkey packages offer network options, which allow more than one computer to access the particle count data. Networking provides a means for high-speed data transmission between computers. Computer data are usually stored in "files," which can be transferred across the network in a fraction of a second. Data collected by SCADA software are also stored in files, which are continuously updated as new data are received.

The particle counting software can be used to create files that are then transferred to the SCADA computer over a network connection. This method is more robust than DDE, as the data are transferred intact and not handed off as they are being updated. For this method to work, the data must be transferred in a file format that is compatible with the SCADA system. While this will usually require some additional programming, it is not as complicated as writing a driver. All the particle counter timing and control functions are handled by the particle counter software.

Particle counting and SCADA system data files are usually stored in relational databases. SQL (Structured Query Language) is a programming language that allows creation of interactive routines between database programs. SQL (often called by the name "sequel") greatly streamlines the process of file sharing between applications, and is supported by the most up-to-date systems on the market.

3. Central Controller Unit

Some particle counter systems can be supplied with a central controller unit, which acts as the "control center" for the system. It provides the necessary timing and control functions, and collects the data from each particle counter. These data are then made available via an RS-232 serial port, which can be accessed by the computer. This type of arrangement may still require a custom driver for direct SCADA interface, but is simpler to implement.

Each of these options has benefits and drawbacks. Before selecting an approach, it is wise to examine all the options available for both the SCADA equipment and the particle counting system. The direct driver interface will usually be more costly in the beginning, but may provide fewer operational problems over the long run. On the other hand, as computer hardware prices continue to drop, it may be more practical to use an extra machine or network to handle the data collection.

The most important thing to remember is that these decisions should always be resolved before specifying and bidding a system. There are way too many pitfalls to leave them up to chance, or to the good intentions of the manufacturers or system suppliers. No one wants a problem system, which will be the source of endless trouble for all parties involved, from the extra hours of technical support to the bad reputation that will accrue, whether justified or not. A poorly thought-out and inefficient system will cost a lot more in time and trouble than the initial expense of doing things correctly from the beginning.

CHAPTER 11

Computerized Data Collection

Most particle counting systems utilize some type of computer interface, either the plant SCADA system or the software and computer provided by the particle counter manufacturers as part of a turnkey system. This chapter covers some of the basics of computers and the computing requirements for particle counting.

A. COMPUTER BASICS

Computers have become such a part of life that it is almost impossible to avoid them. Even those who use them regularly for typing and other office chores may find the many facets of networking and serial data communications overwhelming. This section is intended to provide a brief overview of the basics of computing as it relates to particle counting and data collection.

1. Platforms

All of the turnkey systems provided by the particle counter manufacturers include software designed to run on an IBM Personal Computer (PC) platform. This is the most prevalent form of desktop computer available today. IBM developed the platform, but a seemingly endless variety of "clones" (systems designed by other manufacturers) are available, and are usually less expensive than the IBM brand. The closest competitor is the Macintosh, which has some advantages over the PC, as well as many loyal users, but is not supported by the particle counting manufacturers. For that reason we will focus on the PC in this chapter.

2. Operating Systems

The operating system is the interface between hardware and software, which makes the PC accessible to the outside world. Some SCADA software programs run on operating systems such as UNIX, but the standard turnkey systems are all designed for various versions of Microsoft Windows™ or IBM DOS (Disk Operating System). DOS is rarely used by itself anymore, although it remains a part of the Windows operating system. The Windows operating systems provide a graphic user interface (GUI) as opposed to the text-based DOS. (GUI allows tasks to be initiated by pointing to images on the screen with a mouse as opposed to typing in text commands.)

The earlier versions of Windows (3.1 and 3.11) are designed to process data in 16-bit words, and the more recent Windows 95/98/2000 versions and Windows NT systems are equipped to handle 32-bit words. Simply put, the Windows 95 and NT packages are able to process twice the amount of information during each operation. Several other improvements have been provided in these later operating systems. The most important of these are the ability to run several programs simultaneously with greater reliability, and improved networking and data communications. Windows NT is specifically designed for networked systems, although it can be run as a stand-alone.

3. Processor

The heart of any computer is the "processor." The processor is primarily categorized by the speed at which it performs the computational tasks presented to it. Usually these speeds are measured in terms of the clock frequency at which the processor is operated. Data are processed in a sequential manner, and each step in the computation is initiated by a continuously running "clock." Available technology at the time of writing is providing processor speeds in excess of 800 MHz (megahertz, or million cycles per second).

It must be kept in mind that the processor performs a multitude of functions, which easily comprise millions of steps. It controls a host of "peripheral devices" used for data input, storage, and display. As processor speeds have increased, software has been designed to take advantage of these speeds in every way possible. Simplified user interfaces require a lot of extra processing, since the computer does not operate in a manner that is consistent with normal human thought and reasoning. Each significant increase in processing "power" opens up new areas to be exploited. Some current examples are full-motion video and human speech recognition, both of which require continuous high-speed processing.

4. Memory

Computer memory, known as RAM or random access memory, is temporary storage space for the processor. Unlike recorded media, such as magnetic disks or CD ROMs, the computer memory does not permanently retain data. The data are maintained in integrated circuits as electrical signals that can be accessed and updated much more quickly than a permanent storage device, which requires mechanical access. RAM acts as a "liaison" between the permanent storage media, as well as the input and output devices, and the processor. It is a sort of "on-deck circle" where the data wait for their turn to be processed. This memory also acts as a sort of "scratch pad" for the processor, where interim values are stored during complex operations.

The amount of RAM installed is directly related to the overall speed of the computer. An insufficient amount will become a "bottleneck." When the capacity of the RAM is exceeded, temporary data must be stored on the hard disk drive, a much slower mechanical device. The data stored in RAM are "volatile," i.e., will be retained only as long as electrical power is supplied to the computer. Most of the particle counter data collection packages store the data in RAM for 10 or 15 minutes before writing it to the hard disk. A momentary power outage may result in the loss of this data.

5. Storage Media

Computers are supplied with several types of media for permanent storage of data. "Permanent" is a somewhat misleading term, as these media are somewhat delicate, and will not last forever. Permanent is used to distinguish this type of data storage from volatile forms of storage such as RAM. Permanent media will retain data after the electrical power is turned off. Many forms of permanent storage media are designed for removal and transport of data between machines, as well as for physical storage outside the computer.

a. Hard Disk

The most basic permanent storage device is the hard disk. It is an electrome-chanical device that contains a permanently mounted magnetic disk. The hard disk is designed to remain an integral part of the computer assembly, and is the primary location for all of the software and data used on a regular basis. As mentioned above, it is also the location where data are stored temporarily when the capacity of the computer RAM is exceeded.

Data are stored on the hard disk through a moving "head," which imprints the magnetic disk. Data are constantly added and removed from the hard disk during most operations.

Current hard disk capacities are in the range of several gigabytes (a gigabyte is equal to 1 billion bytes). As mentioned above, advances in speed have led to increasingly demanding software applications. It stands to reason that such applications are large and require increased amounts of data storage capacity.

b. Floppy Diskette

The most familiar form of portable permanent media is the floppy diskette. Floppy diskettes are small magnetic disks encased in a protective plastic shell. They are inexpensive, and can hold about 1.4 MB (million bytes) of data in the standard 3.5-inch format.

The increasing size of software applications and the larger resulting data files have made the small floppy disk less practical. Several disks may be required to load a single program or file. Most programs are now provided on CD ROMs, which contain about 500 times the storage capacity.

c. CD ROM

The CD ROM employs the same type of technology as the audio CD, which stores digitally encoded sound information. The data are structured differently on the CD ROM. ROM is an acronym for read only memory. This is because data can only be read from the CD ROM, not written to it. CD ROMs are primarily used to store software programs, along with manuals or catalogs. A full set of encyclopedia can be stored on a single CD ROM.

Recordable CD ROMs are now available. They are made of a different material, and require a special recording unit. However, they can be read from any standard CD ROM device. They are most useful for small-scale distribution of large amounts of data. Some can only be recorded once, unlike magnetic disks, and are not as practical for day-to-day data backup, although they may be used for long-term archiving. Rewritable versions may be recorded over several times.

d. Other Permanent Storage Media

Several forms of permanent storage media have been developed in the past few years. Most of them are designed for backup of the hard-drive data, or for transporting large amounts of data. Tape backup systems are available up to several gigabytes in size, and are used to back up an entire hard disk. Tape backup is usually performed on a routine or automated basis, to prevent large amounts of data loss because of hard disk failure.

Large-sized floppy disks are available in several proprietary formats, providing storage capacities up to a gigabyte or more. They are most useful for manual data backup and transport. Data can be accessed more quickly than from a tape backup, and these disks can be used as an additional hard-disk drive if necessary. The access time exceeds that of the standard floppy diskette, but is not as fast as a standard hard drive.

Many types of permanent storage media are being developed to meet the increasing demands for larger-capacity data storage and handling.

6. Communications Ports

Standard PCs are equipped with ports for transferring data directly to other computers or devices. The two most common are serial and parallel ports. Additional circuits can be installed to provide network communications. Each of these options is described below:

a. Serial Port

The serial port is designed for transmitting data sequentially, which is the simplest and most common method. This is the type of port used for communicating with particle counters, and is also used for modems (devices that transmit data via telephone lines), as well as many other instruments commonly found in the drinking water treatment plant. Serial data are transmitted at different rates, commonly referred to as the "baud" rate. The speed of communication depends upon the capabilities of the other devices, the length of the communication line, and other factors.

Since serial data are transmitted one bit at a time, only two wires are necessary in most cases. (A third wire is used for a common return or shield.) Some units require additional "hand-shaking" lines for specific signals used to regulate the flow of data between the two units. Serial interfaces are not standardized, and can be somewhat complex.

As mentioned earlier, most particle counting systems communicate with the data collection computer through the serial port. Usually a signal adapter of some sort is required.

b. Parallel Port

The parallel port is used to transmit data in "parallel," i.e., several bits at the same time. This method moves the data more rapidly. The most common use for this port is to send data to the printer. Printouts, particularly of graphic images, contain a large amount of data. Windows 95/98/2000 provide a fairly direct method for transmitting data between two computers via this port. It is often used for transferring files between a portable computer and a desktop computer. Fortunately, parallel port protocols are standardized. Some instruments are designed to send data to the computer via the parallel port, but none is found in the application areas covered in this book.

c. Network Card

Network cards are often provided standard with off-the-shelf computer systems, and are increasingly being used to provide networked connections between computers. They provide much higher speeds of data transfer than serial or parallel ports, and operations carried out over a network will usually appear to be as fast as if they were done on a single computer. There are several network protocols, a discussion of which is well beyond the scope of this book.

d. USB

USB (Universal Serial Bus) has been fully implemented in Windows 98 and later versions. It is a high-speed serial interface designed to allow easy connection of computer peripherals such as printers, scanners, modems, and any number of other devices. This interface was developed to create a fixed standard to clear up the problems often encountered with standard serial and parallel ports. USB is just now beginning to gain widespread popularity, and is not yet supported by the particle counting manufacturers.

7. Additional Components

a. Motherboard

A typical PC is built around what is called the "motherboard." The motherboard is a circuit board that provides the interface between the processor and the RAM, storage media, input and output devices, and any number of optional components. This board is directly connected to the power supply, and routes the proper power signals to all of the attached devices. Some motherboards include built-in data ports and/or video drivers, whereas others require that those items be added as separate boards. Disk and CD ROM drives are connected to the motherboard with multiconductor "ribbon" cables, to provide power and connection to the data bus.

Any number of specialty-type circuit boards have been designed to plug directly into the motherboard. The motherboard contains several board connectors, usually referred to as slots. These "slots" provide direct connection to the power and data bus. Three types of slots are commonly found in IBM-type motherboards. They are known as ISA, VESA, and PCI. Most motherboards contain at least two of these types. ISA boards were designed for a 16-bit data bus, whereas the others can handle a 32-bit bus. Most of the standard PC accessory boards can be found for each of these types of slots. Some older specialty boards may only be available in ISA format, while the more advanced boards will require one of the 32-bit standards. Most particle counting systems do not employ specialized plug-in boards.

RAM is also plugged into the motherboard, in a separate group of connectors. Windows 95 requires at least 16 MB of RAM, and it is possible to expand up to 128 MB or more on some motherboards. RAM is provided on small circuit cards in quantities from 1 up to 64 MB. Usually from four to eight of these RAM modules can be plugged into the motherboard. These modules must be added in pairs of identical sizes. It is advisable to use larger-capacity RAM modules to leave room for future expansion. RAM is provided in several types, and must be matched up properly.

Most motherboards allow for the processor to be replaced and upgraded. When purchasing a computer, take into account the upgrade limits of the motherboard.

b. Mouse and Keyboard

The mouse and keyboard are the means of controlling and inputting data into the computer. Several types of mice and keyboards are available, most providing different "ergonomic" features designed to reduce fatigue. Some mice provide a third button for accessing special features in particular software programs.

The mouse may be interfaced into the computer in one of two ways. The most desirable is the PS/2 port interface. This is a specially designed mouse port with a small round connector. Some systems require that the mouse be connected to one of the serial ports. This will leave the computer with only one available serial port, which will be required for the particle counter signal interface.

The keyboard is available in many shapes and sizes, designed for enhanced ergonomics. A Windows 95 keyboard contains a couple of extra keys, which allow direct access to some of the Windows 95/98/2000 functions.

c. Display

Desktop computers use a cathode ray tube (CRT) display. This display is similar to that of a television, except that it is higher resolution. The CRT display is commonly referred to as a "monitor," and is available in sizes ranging from 14 to 21 inches. These measurements are the same as for televisions, and denote the diagonal size of the screen. Currently available models are designated as SVGA, or super VGA, which refers to the screen resolution.

Resolution is controlled by the video card installed in the computer. Higher-resolution modes require a lot of memory, so most of these cards have slots for adding additional memory. This additional memory allows sophisticated graphic images to be displayed more quickly. Higher-resolution settings allow the displayed images to be reduced in size without losing clarity. More items can be displayed at once, as well as longer time periods for trend graphs, etc.

d. Modem

A modem is a device designed to translate data to and from audible tones so that it can be transmitted over a conventional telephone line. Modems may be installed inside the computer in one of the card "slots," or be externally connected to a serial port. Current technology limits the data transmission rate to 56,000 baud. Some older phone systems will only allow data transmission at about half that rate. Newer high-speed technologies, such as xDSL and cable modems, allow access up to several Megabits per seconds.

B. COMPUTER REQUIREMENTS
FOR PARTICLE COUNTING SYSTEMS

Most of the particle counting systems supplied turnkey from the manufacturer will come complete with a computer, or will specify the minimum requirements for the computer. All the current systems are designed for IBM PC platforms, with Windows operating systems. This is far and away the most commonly used computer platform available. Since most of the commercially available software is written for this platform, additional tools for data presentation and analysis are plentiful.

When purchasing a computer to be used with a turnkey system, always meet or exceed the minimum requirements specified by the manufacturer. In some cases the program may run on a lesser machine, but will run slowly and cause irritation to the user. The following recommendations will help in determining the best computer selection for a particle counting system.

1. Computer Selection Guidelines

The particle counting system will only perform as well as the computer at the heart of the system. When selecting a computer, always try to achieve maximum performance without unnecessary cost. The major advances in computer speed and performance, which have accompanied a dramatic drop in pricing, have made this task much easier. The following guidelines should be helpful for making the correct choices when purchasing a new or upgraded computer for a particle counting system.

a. Purpose

Always keep the purpose of the system foremost in mind. The computer should provide maximum performance in areas critical to particle counter system performance. Features and functions unrelated to particle counting should be minimized. Do not run the office paperwork and accounting functions on the same machine. Discourage any usage unrelated to the task at hand. Computers are too inexpensive to skimp in this area. The more unnecessary work is performed on the computer, the more likely the system will crash and data be lost.

Some of the more expensive features currently revolve around full-motion video and speech recognition, which require a lot of memory and processing power. These advances are unnecessary for particle counting software, and will only add unnecessary costs. Large CRT displays make viewing data more comfortable, but the top-end high-resolution displays are designed for detailed graphic layout and CAD drawing, and are way beyond the requirements of the particle counting system.

Particle counting software has improved a great deal in the last couple of years, but is still pretty far behind the curve in terms of available software technology. It will never be "cutting edge" in that sense, so keep that in mind when choosing a computer system.

b. Performance

A computer is a self-contained system. Processor speed is not the primary consideration, nor is the amount of memory or the size of the hard disk. All of these items work together, and should be kept in proportion. Do not minimize RAM to increase processor speed, as both contribute to the overall speed of the machine.

Processors and hard disks are constantly being improved and made faster and bigger. When pricing a system, look for the "break point" where a good amount of savings can be achieved. If this point occurs at a performance level that easily exceeds the manufacturer's requirements, it is probably a good choice. Do not feel compelled to buy the fastest and most-feature-laden system, when a suitable machine is available at a much lower cost.

Do not be concerned about obsolescence, as that is a part of life where computers are concerned. If the computer has to be upgraded every year or two, that is no big deal. It can always be put to use elsewhere in the plant. Do not buy an overpriced computer based on the manufacturer's promise of a new software package that will be available "soon." By the time "soon" rolls around, several new advances in computing will have become available, and prices will be even lower.

c. Computer Brand

Large metropolitan areas sport dozens of small computer outlets that can provide equipment at a very low cost. Several major vendors sell equipment on a national scale, usually at a higher cost. Any and everything can be found via mail-order catalogs or the Internet. Which route is best?

In most cases, the final answer will come down to support. A small local shop may be able to provide quick and convenient support, especially in a small town. A shop that is well established and has a proven track record is a good choice. Mail-order houses will usually require that the computer be returned to them for repair, which is impractical. The large-scale vendors will usually provide next-day shipment of defective components, which the user can replace and return for warranty credit. If the water plant has a competent technician who is capable of repairing the computers, this can work out well.

Most small shops that build "custom" computers use the cheapest available components, and are constantly switching suppliers as prices fluctuate. The extremely low profit margins make this practice necessary. Most computer components are throw-away items, as the repair costs more than a new part. The "name brands" are only a little more consistent with component selection. Some of them maintain good records of each machine sold, and can quickly access that information. The small shops will not be able to maintain such records, placing the burden on the user to keep track of all the documentation supplied with the computer. The name-brand dealers usually provide better documentation, and often post it on the Internet for easy access.

Particle counting manufacturers will usually supply a name-brand computer with their turnkey systems. Some computers are not compatible with their software, and there is no way to test all the thousands of possible computers that are available. Unlike the water plants, computers do not have to perform according to enforced regulations. While it is in the interest of computer suppliers to make their machines work according to accepted standards, it is not possible to guarantee this. Since particle counting manufacturers ship machines all over the country, it is important to have nationwide support available. The standardization and documentation of the name brands is also important to them.

Unless well-established, reputable computer shops are available locally, the national brands will most likely be the best choice. A few hundred dollars in price difference will be insignificant in the long run. Along with IBM, some of the better brand-name systems include Gateway, Dell, Micron, Compaq, and Hewlett-Packard.

2. Recommended Computer for Particle Counting Systems

With the ever-changing technology in the computer industry, any specifics may well be obsolete by the time this book is printed. This section provides recommended components for a standard particle counting computer without specifying processor speed, hard disk size, etc. Specifics should be determined in consultation with the manufacturer of the system being selected, and according to the guidelines in the previous section.

No attempt is made to extend these recommendations to SCADA system computers or other special systems. Such requirements are beyond the scope of this book.

a. Power Conditioning

Most water treatment plants experience power outages or brownouts on a regular basis. In most cases they last only a few seconds, but that is long enough to cause the computer to shut off and restart. Needless to say, several minutes of data can be lost during this time. In all cases, a UPS (uninterruptible power supply) should be used on the computer. A UPS will provide several minutes of temporary power, which will prevent most of the problems. Longer outages will occur on occasion, but the particle counters will usually be down during these periods as well, so the computer will not be the cause of the data loss. The UPS will provide enough time for the computer to be shut down properly when a prolonged power outage is anticipated.

Surge suppressors should be placed on the power lines for all computer components. These will prevent damage from transient spikes that can occur periodically. In most cases, surge suppressors will not stop transients resulting from a direct lightning strike, but are effective for lesser surges. Modems can be damaged through the telephone line, and should be protected with special telephone line-surge suppressors.

b. Operating System

In most cases, the operating system will be dictated by the software being used. It is best to use the most popular and widely supported system on which the software will run, if multiple options are available. Do not jump on the newest operating system until it has been approved by the particle counting manufacturer. Likewise, do not stick with an outmoded one because of familiarity.

c. Computer Components

Select the processor, memory, and hard disk drive according to the guidelines in the previous selection. Always exceed minimum requirements, which are not established for optimal performance. Stay with popular and commonly available processor types and peripheral standards.

d. Backup

Always back up data, in case of hard disk crash or other problems. Tape backup can be automated, but may interfere with the operation of the software. Check with the particle counting manufacturer for guidelines in this area. Other manual backup options are available, as discussed previously. A second hard disk can be installed to provide a backup as well.

A backup computer is a wise choice, especially for larger systems. It can be used in another capacity in the plant until needed.

e. Support Software

In many cases, a spreadsheet program is a useful companion to the particle counting system. Often the plant will already be using a spreadsheet package of some sort. Any number of useful utility programs can improve efficiency. Exercise caution before loading extra programs that might interfere with the operation of the particle counting software. If possible, perform data manipulation on a different machine to avoid problems with the particle counting system operation.

f. Modem

Some manufacturers offer dial-up support, and can access the plant particle counting system over the telephone. This provides a means for correcting bugs or upgrading the program directly, and can be of special value to smaller plants having operators who are less computer oriented. Modems will usually require direct access to an outside telephone line, and will not operate through many PBX or telephone switching systems.

Modems can provide access to the Internet for downloading software upgrades. Although the plant should be set up for the Internet and e-mail, this is better done on a machine other than the one used for particle counting. The direct telephone line can be shared by the machines, as it will not be often needed for the particle counting computer.

g. Networking

Larger plants may want to take advantage of the benefits of networking. Networking is becoming more and more inexpensive, and is a good way to share data for backup and manipulation. The data collection computer can be left alone, while another machine is used to create reports and analyze the data.

C. DATA MANAGEMENT

Particle counters produce a lot of data, and it is easy to be overwhelmed if an efficient means of managing that data is not employed. The problem is no longer one of storage capacity, as the costs of data storage have plummeted with the improvements in technology. The goal is to have ready access to relevant data.

How much data is "enough"? A number of factors are involved in making that determination, and even more opinions. These decisions relate more specifically to one's overall approach to particle counting. It is important to remember that particle counting came into vogue in the drinking water industry a few years ahead of the computer technology that has greatly simplified data management. Much of the debate about particle counting is still colored by this early experience, and must be taken with a grain of salt. On the other hand, modern life seems to be overburdened with "data" in so many areas, that a reaction against piling on ever greater amounts is certainly justifiable.

The first thing to remember is that data that are destroyed can never be recovered. This is a simple enough concept, but does have a bearing on the subject. It is always possible that new discoveries could bring a whole new outlook on what would seem to be routine information. It may be of value to retrace the data from several years back. As storage becomes less and less costly, there is less of a reason to discard data. If a year's worth of particle counting data can be stored for a few dollars, is it worth the cost?

The issue of data management is more properly centered on efficiency of access than storage. The data must be maintained in a systematic manner which allows the data to be retrieved and manipulated easily. In most cases, the data should be stored according to the date on which it was collected. This is the most logical way to store data, as most of the events prompting the review of old data will be related to seasonal changes, or as a result of events that have come to light after the fact. For example, a water system that has experienced a *Cryptosporidium* outbreak may want to review all the plant data for several weeks or months prior to the date of discovery, to determine if any signs of the problem could be detected.

The ease of access will also involve the structure of the data, and whether the data can be retrieved by the particle counting software only, or are accessible to other programs. This is an important point, because software updates may not be compatible with older data file formats. If the data are to be sent to another site, or shared with other utilities or regulatory agencies, a universal file format will simplify this transfer.

In addition to these concerns about long-term data management, there are issues involving data structure for everyday use. These will be briefly covered below.

1. Reporting

Reports are necessary not only for meeting regulatory requirements, but are important summaries of the plant treatment performance for the day, week, or month that they cover. They must be compact but informative enough to transmit an accurate picture of the particle counting data. The report should provide a brief outline of the average operating conditions as well as any odd or unusual occurrences. Enough information should be presented to refer the operator back to the appropriate data files for further study. An anomaly discovered should be characterized such that it can be easily referenced when a similar event occurs.

Whenever possible, the report should contain a record of any process changes or other occurrences that can have an effect on the particle count data.

Any number of report formats may be used. Some of the particle counting software programs provide user customizable reports. These provide a means for operators to create reports that are adapted to their particular application. Reports should be structured in a way to present the data in a meaningful manner, and not just to fit the mold of the other plant data. The hourly or daily "minimums, maximums, and averages" are useful, but do not tell the whole story. Where in the course of the filter run do these peaks and troughs occur? Were they due to fluctuations in filter loading rates, or to chemical dosage? There are all sorts of factors that must be figured into the proper interpretation of particle count data. While all of them cannot be included in every report, enough information should be provided to point to the most critical.

As discussed in Part I, the trend graph provides the most complete and accessible form of data presentation. A graphical presentation of the most recent filter run compared with a running average of all the filter runs for the year might be quite useful. The particle counting system can open up a lot of new ways of looking at familiar data, especially when few regulations have been established to force one into a mold.

D. UPGRADING EQUIPMENT AND SOFTWARE

One of the most important by-products of the fast-moving changes in the computer industry is the need for what seems like continual upgrades of software and computer equipment. This is less of a problem in the particle counting industry than in the world of commercial computer equipment, because of the relatively small size of the particle counting industry. Competition is not as fierce, and, until recently, there was so little competition among particle counting companies in the drinking water industry that little effort was expended beyond the bare minimum required to make the systems work. That has changed as more companies have moved into the growing particle counter market. It has taken these companies a while to catch on to the fact that the software is what puts the "pizzazz" into particle counting. The software is what the operator sees and interacts with, so it will naturally be the most appealing part of the system.

Now that the importance of the software presentation has been impressed upon the managers and bean counters of the particle counting companies, a new attitude has emerged. Software has gone from being a "necessary evil" to the star of the parade. This has been a benefit to the drinking water operator because we are now beginning to see some nicely designed and useful software for online particle counting. (Grab samplers are another story.) The downside of this change in outlook is that the familiar "upgrade" treadmill may become rampant in the particle counting industry. Software is much easier to change than hardware, so it is only natural that this route be followed. This will not be a big problem as long as the end user is well aware of what matters and what does not, and can calmly weigh the options.

Another reason for continual software upgrades is that there is no software that is "bug"-free. Usually software is rushed out to market already behind schedule, and bugs are worked out as they are discovered. This is understandable, since the many possibilities for software application cannot all be tested before the product is finished. The same holds true for particle counting software. Although the pace is somewhat slower, it follows the same trajectory.

Much of the older particle counting equipment on the market will have to be upgraded to take advantage of the better software packages available. Some of this earlier particle counting equipment was designed before computerized data collection was much more than an afterthought. In most cases, the biggest changes have come in the counting electronics and communications circuitry. The particle sensors have not changed a great deal, except perhaps for mechanical changes designed to lower manufacturing costs. Many of the earlier sensors were designed for more-demanding industrial applications, and are more rugged and reliable than the newer versions. They will become obsolete when the calibration and repair support for them is discontinued.

Most of the particle counting equipment will be upgraded *in toto*, for the reasons mentioned above. There is little reason to stay with the same manufacturer other than the cost break that may be provided for the upgrade. It is likely that competing manufacturers will provide similar incentives, so there is no need to limit one's options. If the manufacturer of the older equipment has provided excellent service and support from the beginning, there is a good reason to stick with them. Otherwise, work out the best deal possible, both in terms of price and performance.

E. NETWORKING AND REMOTE COMMUNICATIONS

Networking has been discussed earlier as a method of expanding access to the particle count data as well as an efficient means of SCADA interface. A network may comprise only two or three computers, or involve a whole utility. The many available options are beyond the scope of this book.

Remote communications can be defined as retrieving data from instruments mounted in locations remote from the water plant, such as in pump stations or water towers. There has been little interest in placing particle counters in such locations, and there is no important application for them out in the distribution system. In most cases, remote applications fall under the larger heading of SCADA integration.

CHAPTER 12

Putting It All Together

All of the elements covered in Chapters 6 through 11 must be combined to form a complete particle counting system. In most cases, the system will be supplied by a single manufacturer, as there is no way to pick and choose the best of what each has to offer and build the system from scratch. On the plus side, much of the selection process is simplified as a result of this. This chapter presents some of the considerations for selecting the correct system for a given application. Obviously, there is no way to cover every possible situation, so only general guidelines can be presented.

Among these considerations are the size of the treatment plant, the training and capabilities of the operating staff, the possibility of future plant expansion, and the financial situation, as well as the technical requirements for the method of treatment employed. It is important to note the track records of the various manufacturers with other installations in the area, and their willingness and ability to provide service and support. A certain amount of subjective feel for the way a particular system operates will also come into play.

A. THE TREATMENT PLANT

1. Size and Future Plans

The size of the treatment plant will be a major factor in determining the most suitable system. How many points will be monitored? Will the complete system be installed at one time, or will it be expanded gradually? Will the plant be expanded in the future, or will it be replaced in a few years?

Small plants that will monitor only three or four points may consider a 4 to 20 mA output system, whereas larger plants should not. Smaller plants, or those planning to gradually build up a larger system, will usually find that the turnkey systems provided by the manufacturers are the simplest option. The same would hold true for plants scheduled to be replaced in a few years. The turnkey system can be moved to another site with relative ease. Large plants preparing to install a lot of particle counters at one time should look seriously at some sort of SCADA integration.

If a SCADA system is scheduled for installation in the next year or two, the turnkey system can be a good interim approach. It will provide the opportunity to learn particle counting before having to run the entire SCADA system. With some advance preparation, the particle counting system can be integrated into the new SCADA system without a lot of trouble. In such cases, the manufacturer's willingness to provide support is critical.

2. Staff

The capabilities of the operating staff are always a prime concern. Many operators are willing but do not have the training necessary to handle a particle counting system. Others do not want to expend the effort even if the opportunity is presented. It is important to make a realistic appraisal of the staff's attitudes, capabilities, and opportunities for training before selecting a system. It is much better to have a simple system that provides some useful information than a sophisticated one collecting dust.

Just as important as the willingness to learn and operate the particle counting system is the efficiency of the operations and maintenance staff. Is the staff going to keep up the routine maintenance without a lot of prompting, or will it take a few alarms to produce any action? How well are the current plant instruments maintained? Much of the demand for electronic flowmeters, which are costly and often do not work too well, is driven by the desire to compensate for poor maintenance. These extra components only increase the need for good maintenance instead of reducing it.

3. Treatment Process

The type of treatment process employed may have a bearing on the type of particle counting system selected. Conventional plants with particle counters monitoring raw water will require evaluation of online dilution systems as well as the true coincidence limits of the available particle sensors. For most applications, the physical location of the sample points will have more bearing than the treatment process employed. It may be that the ease of maintenance or the manufacturer's ability to provide assistance in solving difficult sample delivery problems becomes the deciding factor.

B. EQUIPMENT FEATURES

Prioritizing the many features available can be difficult. Some of them are of real value, while others involve mere "specmanship." Except for electronic flowmeters, few of the various features add significant cost to the system. Most are an integral part of the standard product line.

1. Packaging

The packaging of the particle counters is not for attractiveness, as the units are usually located in out-of-the-way places where few will ever see them. Fancy packaging provides a "high-tech" look for trade shows and advertisements, but is of little importance from an operational standpoint.

The important consideration is that the packaging provides proper protection for the instruments and easy access for installation and maintenance. For the most part, standard NEMA 4 or 4X enclosures are provided. Of course, once holes are drilled in these enclosures they are technically no longer NEMA rated. This is unavoidable, and is typical for most drinking water process instruments. Some of the earlier particle counters were not packaged properly for water plant environments, but the manufacturers have learned a good deal in the last few years. That said, many of them still run the sample into the NEMA enclosure, providing the greatest potential for leaks and hazards.

Passing the sample into the enclosure means that the enclosure must be opened for the sensor to be cleaned. As cleaning involves removing and reconnecting the sample inlet tubing, the potential for problems is greatly increased. It is not uncommon to find plants where a particle counter has been filled up with water because of a damaged or leaking fitting.

When water is passed into the enclosure, the best way to reduce safety hazards is to place the electrical power supply in a separate enclosure. This increases the amount of installation required, as a second enclosure must be mounted and wired to the particle counter. On the other hand, removal of the particle counter for service or calibration is simplified if the AC power does not have to be handled.

When the enclosure must be opened for cleaning the sensor, the time and effort required is increased. This makes the task of maintaining the sensor a little harder, and thus easier to neglect. If any circuitry is exposed, there is always the likelihood of water being spilled on it during the cleaning. As anyone who has worked with water and instruments can tell you, water always seems to get on any and everything.

Particle counters do have to be removed periodically, whether for calibration or service. If auxiliary inputs are connected inside the enclosure, a good deal of effort may be required to remove and reinstall the unit. The instrument technician will have to be involved, to prevent miswiring which could damage the particle counter or other equipment.

The physical dimensions of the enclosure are usually not a problem from the standpoint of mounting space. Most filter galleries provide plenty of room to mount instruments. The size of the enclosure is more important from the standpoint of access to wiring terminals, which will simplify installation and removal.

2. Sensor Characteristics

The particle counting sensor is the heart of the system. In addition to providing the particle pulse data, it is the interface with the sample stream. It must be evaluated in terms of both performance and mechanical structure, which are interrelated.

a. Flow Cell

The flow cell must be rugged enough to maintain both its mechanical and optical integrity through years of continuous sample flow with periodic cleaning and clog removal. The size of the flow cell orifice is proportional to the ease of maintenance, but inversely proportional to resolution and coincidence, so a trade-off is necessary. Some flow cells are virtually indestructible, whereas others are made of more fragile materials. The most fragile can be replaced in the field, whereas the tougher ones cannot, but should not need replacement. In any case, the particle counter is only as good as the flow cell. If it is damaged or not properly maintained, the rest of the instrument is useless.

Once again, maintenance is a key consideration. Will the more delicate flow cells be handled properly, or will a careless, ham-fisted operator tear them up repeatedly? A sensor having poorer resolution or a lower concentration limit but which will stay in operation is preferable to one with better specs that is not maintained properly.

b. Sensor Performance

The more sensitive the particle counting system, the better. While many people joke about choosing the particle counter "that counts the lowest," in reality, the opposite should be preferred. The distinguishing feature of the particle counter is sensitivity to small amounts of tiny particles. Any regulations that provide incentive to minimize this sensitivity will be counterproductive.

Resolution is important for count matching and will be of great concern if and when sizing calibration standards are set for the industry. Coincidence is important as well, especially if log removals are (unwisely) made into regulatory standards.

Sensor performance must be carefully considered in light of these factors. Much will depend on the operator's willingness to use particle counting as a process tool as opposed to a "by the book" way to meet regulations. For now, this is not an issue, but will be at some time in the future.

3. Counter Features

Which counting features are of most practical value? How do we plan for the future, when regulations have still not been determined? For the most part, these decisions must be based on the particular application. Once again, if there is a willingness to use the particle counters as process optimization tools, the clouded future of the regulatory situation will carry less weight.

a. User-Selectable Size Ranges

In most cases, it is hard to go wrong with this option. If regulations requiring specific size ranges are implemented, it is a simple matter to change them. This feature is also quite useful for the more ambitious operator who would like to experiment with different size information. Outside of experimental or research applications, size ranges are not changed often. Fixed size range units can be reset by the manufacturers, so perhaps once or twice in the life of the particle counting system, this extra expense may be incurred.

b. Type of Counting Circuitry

The more-advanced AD counters provide a better platform for count matching and more precise calibration than the simpler comparator-type circuits. More and more specifications are calling for count matching within a given percentage, placing a greater burden on the manufacturers. On the other hand, many are pushing for simplified one- or two-channel counters, in an effort to make them easier to understand and operate. In this case, the more-sophisticated circuitry would be overkill.

c. Auxiliary Inputs

We highly recommend trending turbidity along with the particle count data, as it provides a familiar reference point for those unacquainted with particle counting. A backwash signal input is extremely useful as well, as it eliminates the need to record such information manually. Other inputs such as loss-of-head and filter flow rate may be useful as well.

4. Flow Regulation

A constant-head flow regulator is indispensable. Combined with a low-flow alarm, it should be all that is necessary in most applications. Flowmetering is recommended only if the problems cannot be solved with the constant-head weir, and are not due to poor maintenance. With the exception of the Tritech meter discussed in Chapter 3, the added expense of the flowmeters is not worth the results achieved. The money would be better spent on more important features.

Almost all the problems related to flow regulation will be due to insufficient head. The ability to program the particle counter to work at a lower flow rate provides a solution to this problem. It also allows the sensor to be used on higher-concentration waters. Higher-concentration sources will clog up the sensor flow cell, and only good maintenance will minimize these problems.

5. Data Collection and Presentation

Computerized data collection is recommended in almost every case. The data presentation should be easy to access and interpret. The best presentation will include other relevant plant parameters, whether through the SCADA system or the manufacturer's software package.

a. Trend Display

As emphasized in Part I, trend data are by far the most useful. Any presentation should include readily accessible trend graphs, which can be time- and value-scaled with ease. An important feature is the ability to select various trend values to present on the same graph. In this way, various filter effluents can be compared, or auxiliary data compared with particle count data. It is also useful to have the ability to select a precise time window, so that odd or unexpected events can be studied in detail. If historical and current data can be trended together, comparisons with past events can be made readily.

b. Alarm Display

Properly designed particle counting systems provide alarm functions for flow cell cleaning and low flow, as well as communications problems, which can signal instrument failure. These alarms must be quickly and clearly displayed to provide the opportunity to correct problems before a lot of data is lost. The alarms should also be logged for future reference.

c. Reporting

Flexible and easy-to-configure reports are a must for particle counting systems. Those familiar with commercial spreadsheet and database programs may prefer to use them for reporting, but it is easier for the novice to work within a single program.

d. Historical Data

The system should provide a clean and efficient way to store data. Automatic archiving and backup are desirable features. It is important that data be easily accessible. File structure should allow data to be imported directly into database or spreadsheet programs.

The historical data should be available for access via the main data collection software. The task is made simpler if the data can be accessed using the same trending and reporting commands employed for the current data.

CHAPTER 13

Grab Samplers

Grab samplers are covered in a separate chapter because they operate in a manner substantially different from online systems. The models currently in use are designed for portability, although most of the time they are operated in the drinking water plant laboratory.

A. EQUIPMENT FEATURES

For the most part, the components of the grab-sampling particle counter are identical to those of the online system. The distinguishing feature is the addition of a sample delivery system designed to draw the sample through the particle sensor. The packaging and software must be suited to the different applications requirements. This section presents an overview of these additional features, as well as any considerations regarding similar features used for a different purpose.

One of the most useful features is the ability to function as an online unit. In this mode, a grab sampler can be used as a temporary replacement for an online unit that must be taken out of service, or in a sample location where a permanent online unit has not been installed.

1. Sample Delivery System

a. Portable Grab Samplers

As described in Part I, the grab sampler employs a pump to pull the sample through the particle counter flow cell. In the units currently available on the market, a small gear pump is used. This pump is small and lightweight, and produces a virtually pulseless sample flow. It draws very little power, allowing for battery operation. It is a fairly quiet running unit, compared with the peristaltic pumps commonly found in laboratory applications.

These small gear pumps are not designed for continuous operation, and have an operating life of only a few hundred hours. This will provide several years of operation with grab samples, but precludes long term online usage.

The sample passes through the gear mechanism, and large particles or concentrations of particles can jam the pump. This is not usually a problem when low-concentration samples (concentrations not much higher than the coincidence limit of the sensor) are used. A small strainer should be placed between the sensor outlet and the pump inlet to keep large particles from creating problems.

As described in Part I, this type of sampler does not meter-out the sample by volume, but runs for a fixed time period at a constant flow rate. The sample flow rate is controlled by the voltage applied to the pump. In most cases, the flow rate is set at a fixed level, and is not adjustable by the user. The flow rate should be verified periodically to ensure that it remains constant. As long as the flow remains constant, the data can be normalized to adjust for any deviation from the designed flow rate.

b. Pressurized Batch Samplers

The larger pressurized batch samplers are rarely purchased for water treatment applications now that the much less expensive portable units are available. Several of them are still in use in drinking water plants, so a brief description of the sample delivery mechanism is in order.

The pressurized batch sampler was designed to accommodate viscous oils and lubricants, which require significantly higher pressures to achieve usable flow rates than does water. The sample beaker is placed in a sealed chamber, which is then pressurized with compressed air that has been filtered with a sub-micron filter to prevent the introduction of unwanted particles into the sample. The pressure forces the sample up into a stainless steel tube, which is connected to the inlet of the particle sensor. The sample passes through the particle sensor, and into a collection chamber that measures out the volume of sample.

The volume is metered using a graduated cylinder that is fitted with some type of level-detecting switch. Two switches may be employed, one to initiate the particle count cycle, and one to end it when the desired volume has passed through the particle sensor. The first switch can be positioned to allow some of the sample to flush out any residual particles in the collection tube or sensor flow cell before the actual counts are initiated. Once the run is complete, the sample is then flushed out of the graduated cylinder.

The pressurized batch sampler is not dependent upon a fixed flow rate, but only a fixed volume. The flow rate must remain within the allowable limits of the particle sensor. The flow rate is usually controlled with some type of needle valve adjustment downstream of the sensor.

These batch samplers are far more mechanically sophisticated than the portable grab samplers, and are several times more expensive. There is no reason for them to be purchased for drinking water applications. They are built for heavy-duty service, and many of the units purchased years ago will remain serviceable for years to come.

2. Packaging

As the grab samplers are designed for portability, most are packaged in some sort of carrying case, or equipped with a handle. Some sort of instrument case is used for mounting the components, as well as providing space for extra sample tubing and other accessories.

A primary consideration is that of keeping all the electronic components isolated from any water. It is easy to spill water when performing grab samples, and residual liquid will be left in the tubing after the samples have been run. Access to the sensor for cleaning is also important, as the flow cell windows are more likely to need cleaning with intermittent use. If the window surfaces dry off, it is possible for water spots to form, affecting the optical clarity of the cell. With many different types of samples being run through the same sensor, it becomes easier to contaminate the cell windows. The tubing may need to be replaced more often. As was the case for the online units, cell cleaning involves removing and reinstalling sample tubing to the sensor, which increases the likelihood of leaks.

The pump, strainer, and battery are additional components that will require access for service and maintenance.

3. Counting Electronics

The same types of counting electronics found in online units are used for grab samplers. The different applications for the grab sampler should be taken into account when selecting the type and features of the counting electronics. For example, user-selectable size ranges may be useful, as grab samplers lend themselves to a wider range of experimental applications. Additional size channels may be important for the same reason.

Some of the user-selectable features may require a computer to configure. This is not a problem since a computer should be used to download the data from the unit anyway. An integral keypad and display will be of more value because data can be collected from various sites without connecting the computer, for downloading at a later time. A paper printout may be useful for backup, but should not be the primary means of data storage.

4. Computer Interface

a. Hardware

Most data communications will be done via a standard RS-232 serial port, which is directly connected to a standard PC. Some units also provide an RS-485 interface, which allows the grab sampler to be connected to an online system.

b. Software

Grab-sampling software may be used to upload configuration parameters as well as to download the data collected by the grab sampler. The most that can be expected from the software is that it can be used to organize the data in a usable manner, and provide a file structure that can be imported into commercially available spreadsheet and database programs.

CHAPTER 14

Calibration

Particle counter calibration is a subject which could fill a book in itself. Calibration issues have become more important as particle counting has gained wider acceptance in the drinking water treatment industry. The desire for standards for both instrument accuracy and cross-comparison of data has placed an increasing amount of scrutiny on particle counter calibration as currently practiced.

A. CALIBRATION: AN INEXACT SCIENCE

The foremost thing to keep in mind with regard to particle counter calibration is that it is at best an inexact science. There are several reasons this is so. The first is the nature of the particles being measured with the particle counter. Particles in the range of 1 to 2 μm are difficult enough to measure, let alone produce accurate standards for. Particle counters are pushed to the limits of sensitivity just to measure these particles at all, unlike turbidimeters, which are usually calibrated at a concentration well into the middle of their operating range. The high sensitivity of the particle counter to tiny, microscopic particles makes it difficult to avoid contamination. At present, only size calibration is performed as a standard practice. The standards used are polystyrene latex (PSL) beads, which can be produced to within a few percent of the desired size. Some of the drawbacks to these materials are that they are perfectly spherical, unlike the actual particles found in the water supply, and are composed of a substance not found in natural water sources. On the whole, these problems are not insurmountable, and not even frustrating, if one keeps in mind the value of particle counting as a water quality optimization tool.

B. CALIBRATION MATERIALS

Sizing standards currently in use in drinking water particle counting are PSL spheres. The most prominent supplier of these standards for drinking water treatment is Duke Scientific Corp. of Palo Alto, CA. These spheres are supplied in sizes ranging from less than 1 up to 100 μm and higher. Each size of spheres is supplied in concentrated solution, which is diluted upon use. As mentioned above, these spheres are not made of naturally occurring materials, and have a translucent quality that can affect how they are measured by the particle counter. They are also perfectly round (well, almost), again unlike particles naturally occurring in water. While not perfectly suited for the application, they provide a constant, high-quality, and reasonably affordable standard for size calibration of particle counters.

C. PARTICLE SIZE CALIBRATION

To calibrate a particle counter for sizing accuracy, standards covering a range of sizes are passed through the particle counter one size at a time. These sizes should include the lower sensitivity of the particle counter (usually 2 μm) and 9 or 10 more sizes covering the full sizing range of the particle counter. These standards are introduced into a "particle-free" solution and fed through the particle counter sensor at a constant flow rate. This flow rate should be close to the actual working flow rate of the particle counter (typically 100 ml/minute). Since counts are not being measured, the exact flow rate is not critical.

Figure 14.1 shows a typical calibration setup. The calibration water is passed through a submicron filter in a continuously circulating loop. The filter will remove all particles above 1μm or so after a few passes.

The calibration standards are introduced downstream of the filter, and pass directly through the particle counter sensor before passing through the pump and into the filter. The output of the sensor is monitored with an oscilloscope or PHA. In most cases, the counter electronics are not used for size calibration. They are calibrated independently.

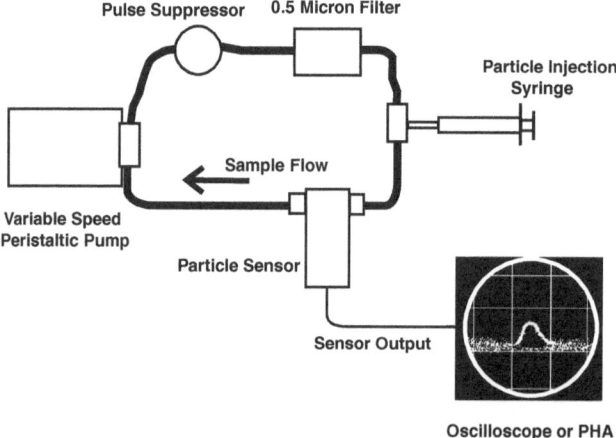

Figure 14.1   Calibration system diagram.

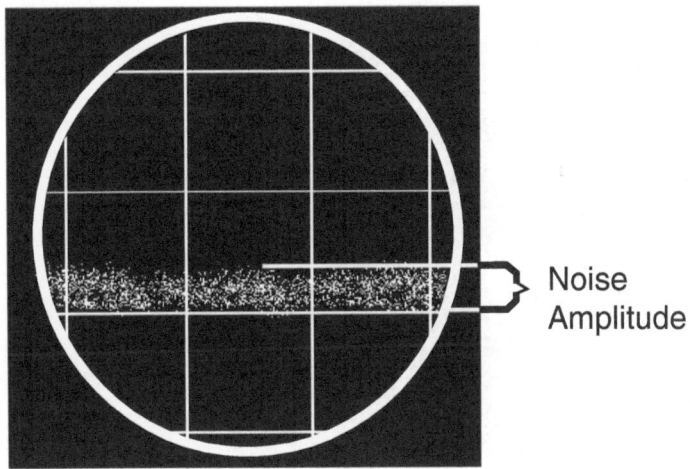

Figure 14.2   Noise measurement.

After the calibration solution is sufficiently cleaned up, the electronic noise level can be recorded; see Figure 14.2. The smallest particle must be at least twice the amplitude of this noise level.

Once the noise level has been recorded, the first particle size is introduced. The concentration level of the particles should reach a level of a few thousand particles per milliliter at the smallest sizes. On an oscilloscope, a cleanly triggered pulse output should produce a clear, bright pulse as in Figure 14.3. If a PHA is in use, it should be restarted when the counts reach a level of 2000 or 3000 ml, to avoid concentration error. A typical PHA output is shown in Figure 14.4.

The advantage of the PHA over an oscilloscope for particle counter sensor calibration should be obvious. When an oscilloscope is triggered, the pulse on the display appears to be still, but is actually a succession of pulses of similar amplitude. The brightness of that pulse occurs because it is actually several pulses "stacked" on top of each other. The amplitude of the pulse is chosen by "eyeballing" the top of the brightly displayed pulse, and reading its value from the oscilloscope grid. The PHA, on the other hand, provides a complete histogram of the pulses, allowing the amplitude to be determined with much greater accuracy.

A further look at Figure 14.4 gives an indication of how accurately the particle counter sizes particles of the same size. The width of the distribution relates to the resolution of the sensor. It should be obvious that any type of consistent calibration standard will have to be built around a PHA.

After the first size standard has been introduced and the data recorded, the system should be allowed to clean up before the next size is introduced. The same steps are repeated for each size standard, usually in order of increasing size. The larger particles will be available in much smaller concentrations. The difficulty with the larger sizes is passing enough particles through the sensor to get a consistent output.

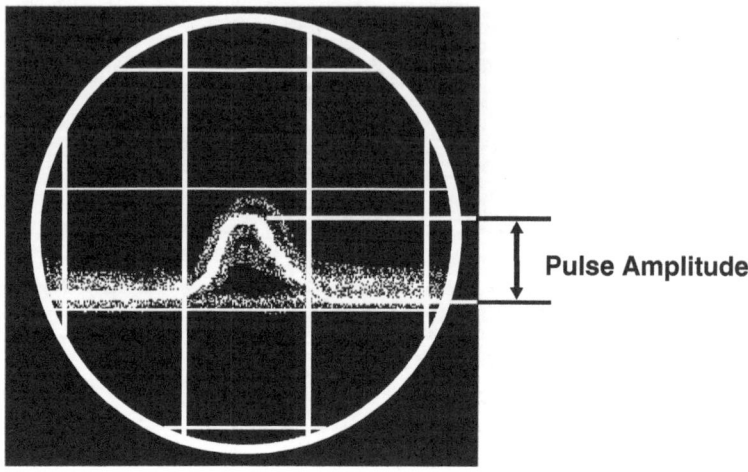

Figure 14.3   Particle size measurement.

D. CALIBRATION CURVE

Once all the particle standards have been run through the sensor, the data can be plotted to produce a calibration curve. A typical curve is shown in Figure 14.5. The particle size is plotted vs. the voltage pulse height (usually in millivolts). In order for this curve to be usable, it must be monotonic (i.e., the voltage level always increases as particle size increases). This is because it would be impossible to distinguish between two different particle sizes if they produced the same pulse amplitude. In addition to producing a calibration curve, a table of values is interpolated to provide the pulse amplitude for each micron increase in particle size. These data are used to program or manually configure the particle size thresholds for the particle counter. The sensor noise level is usually recorded as well.

E. COUNT MATCHING

Some work has been done in the area of count matching particle counters. This involves adjusting the thresholds on two or more counters until they count within a few percent of each other on the same solution. Count matching can be performed manually by physically adjusting individual size thresholds until the desired results are achieved, or automatically using PHA electronics programmed with special algorithms. This latter method has been used on an experimental basis. The particle counters are not necessarily matched to count and size particles correctly, but to count and size them in an identical manner. This is an improvement over existing conditions. At present, there is still no easy way to ensure that particle counters will count identically. The only realistic way to achieve consistent count correlation is

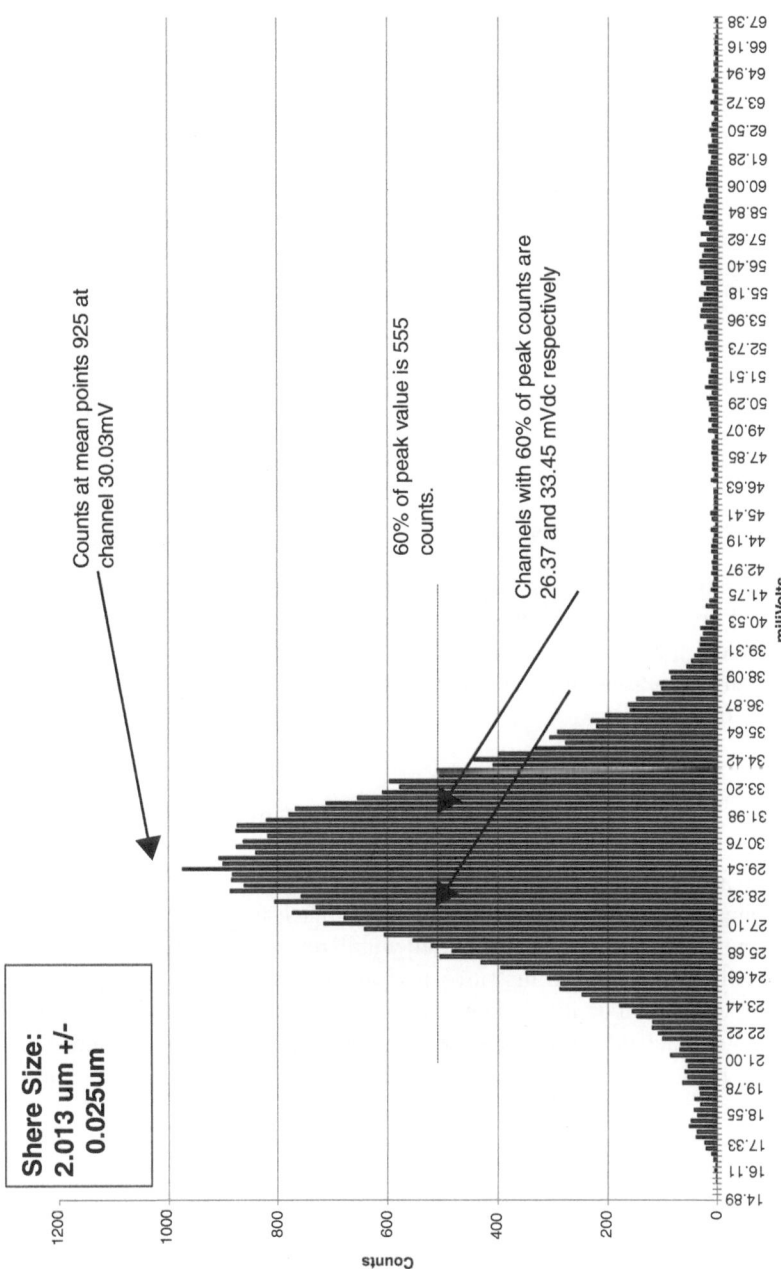

Figure 14.4 PHA calibration data. (Courtesy of Pacific Scientific Instruments, Grants Pass, OR.)

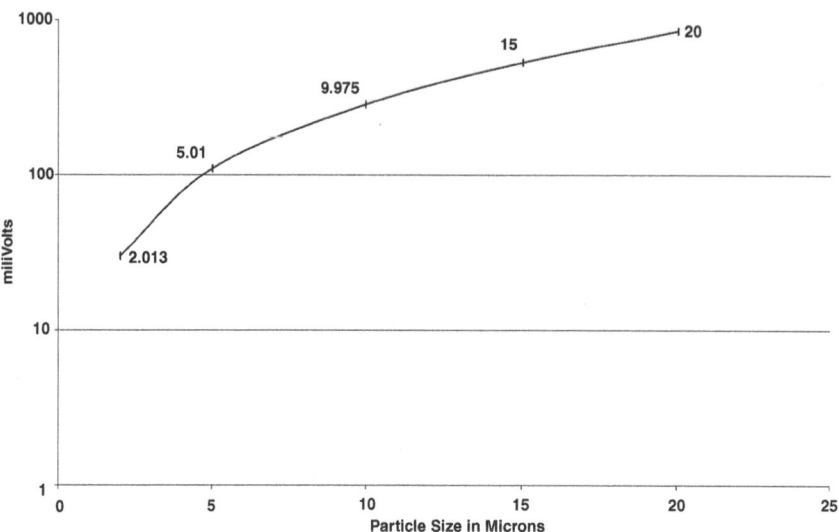

Figure 14.5 Calibration curve.

through improvement in the design and manufacture of the instruments. As this is the most time-consuming and costly approach, it will be the last one taken. There is still room for debate on how important count correlation is to particle counting. Although it would be of great value, it may not be worth a significant increase in the cost of the units.

F. FIELD VERIFICATION OF CALIBRATION

One area of increasing relevance is that of field verification. Although the cost and complexity of in-house calibration is too much for the average drinking water treatment plant, field verification is well within reach. Two types of field verification are readily available. The first involves passing PSL spheres through the sensors while operating online in the plant. The second involves running a grab sampler as a calibrated standard. Currently in development is a count and size standard solution that can be used to verify operation of the particle counter for both counting and sizing accuracy. A brief description of each method is outlined below.

1. Size Verification

This test involves the introduction of PSL spheres in a similar manner to that discussed in Section C. A diluted sample of PSL spheres is introduced into the particle counter, and the output data is checked to verify that the counts are displayed in the appropriate size ranges. Two different-sized particles are enough to verify the calibration of the particle counter. Several methods of varying complexity may be used to perform this test. The various options include different methods of introducing the particles as well as of measuring the response of the particle counter.

a. Particle Sample Introduction

One simple way to introduce the sample is to prepare a small tube with a leurlock-type valve fitting. This allows for a small syringe containing the particle mixture to be connected to the sample inlet of the particle counter. The particles can be injected directly into the filter effluent or finished water stream, as long as the concentration is well above the nominal particle count value for that sample. This method will not work for higher concentration waters. If a filter effluent sample is located near one of these higher concentration sources, it can be used as the source water, after allowing for sufficient clean up time. A sample beaker or similar-type container can be used to collect filter effluent water to be siphoned through the particle counter sensor. A small peristaltic or metering pump can be used to draw the sample through the sensor as well. Particles are mixed into the sample beaker and then drawn through the sensor. In cases where this method is employed, the guidelines for grab-sample handling presented in Chapter 5 should be observed. A precise flow rate is not crucial for a sizing test, as long as it is kept within the manufacturer's recommended range.

b. Grab-Sampler Comparison

If the plant owns a grab sampler, it can be sent out for calibration or verified using the method described above. Once it has been checked, it can be connected in series with the online particle counter to provide verification. In this case, it is not necessary to introduce particles. If this method is used to test clean filter effluent water, particles may need to be added to achieve a sufficient concentration. These particles do not have to be PSL spheres, since the sizing information will be verified by the grab sampler. However, PSL spheres will provide a double-check of the sizing accuracy.

2. Displaying Count Data

The simplest way to display the count data is to use the existing data collection system. Particle size standards should be chosen so that they fall in the center of the size range. For example, if the lowest range is set for 2 to 5 μm, use a 3 μm sample. If the size ranges are user selectable, they can be reset to produce the desired range. It is also acceptable to introduce a particle size that corresponds to a threshold value. If 5 μm PSL spheres are introduced into a particle counter with ranges set at 2 to 5 μm and 5 to 10 μm a roughly equal number of counts should appear in each range. In most cases a size close to the minimum sensitivity (but not equal to it) and a larger size should be used. For a 2 μm particle counter, use 3 and 10 μm PSL spheres. The spheres should be introduced in separate samples, and in concentrations well below the specified coincidence limit (25% or so should be safe).

3. Count Verification

At the time this book is being written, Duke Scientific Corp. is developing a count verification standard. Preliminary tests using 500-ml samples of particle concentrations of 3 and 10 μm particles mixed together at concentrations of 2000 and 200/ml have produced some good results. The samples are pulled through the particle counter with a metering pump calibrated to a 100 ml/minute flow rate. In this test, flow must be accurate to produce the proper count data. The data are collected on the particle counter data acquisition system as described above. The success of this project will depend on producing a consistent sample, which can be run on site without complication. The samples are not inexpensive, so little waste is tolerable.

G. SOME UNRESOLVED ISSUES

Calibration will remain an issue of contention for some time. The sizing and counting accuracy of the instruments will be an unknown until such time as an acceptable standard is developed. Counting accuracy is of more importance than sizing accuracy, but both are interrelated. As particle counter technology filters down to smaller drinking water treatment plants, the tendency will be to move to simpler and simpler systems. These may be one- or two-channel particle counters. In this event, counting accuracy will be the only concern, other than proper size recognition at the low-end sensitivity.

As stated before, the current problems with particle counter calibration should not keep anyone from collecting valuable information from the technology. It does provide a convenient excuse for those unwilling to put forth the effort to use particle counters properly. It also allows the manufacturers to get away with substandard performance in some cases. As with any cutting-edge technology, these issues will be sorted out over time.

PART III

Assessing the Equipment

The growth of particle counting in the drinking water treatment industry has brought about many changes in the technology in the past few years. The first particle counters sold to water plants were modified laboratory units, which worked acceptably but were not tailored to the specific requirements of the application. As the initial trickle of interest swelled into a flood, more attention was directed toward developing suitable equipment.

At present, there are five manufacturers offering standard light-blocking particle counters and systems to the drinking water industry. This part of the book takes a comparative look at the main equipment offerings of four of them. The fifth, ATI, offers only a single-channel particle alarm. The intent is not to present a "good, better, best" rating system, but to present the relevant information in a readily accessible format to allow the reader to make assessments for each given application, with the information from the first two parts of this book as a guide. Any subjective observations about the utility of certain features or assessments regarding ease of use should be taken as opinions of the author.

Specifications are taken from the manufacturers' published literature, and no endorsement of their accuracy is implied. We heartily recommend that information be obtained from the manufacturers before any purchasing decisions are made, both for the reason that the information presented here will be out-of-date sooner or later, and because some of the specific features of a given product may have a rationale not covered by the material presented here. This book is not intended to provide an easy way out of the complicated task of specifying or buying a particle counting system. It is merely intended to leave the reader without excuse for an ill-informed decision.

Actual performance testing is beyond the scope of this book. Studies in that area are being performed, and some of them will be referenced in Appendix 2.

A listing of the current manufacturers is provided in Appendix 1. To account for some of the equipment still found in water plants, which may not be mentioned in this book, a thumbnail sketch of the history of the particle counting manufacturers involved in the water treatment industry follows.

The first particle counters were used in water treatment for research in the late 1970s, and were made by a company named Hiac Royco. These instruments predated the invention of the laser diode, and employed an incandescent white light source.

Little was done in drinking water with particle counters until the late 1980s when a major *Cryptosporidium* outbreak in Carrolton, Georgia, prompted the state health department to look into particle counting as a means of monitoring filters for preventing similar occurrences. By this time, laser diode technology had been introduced, making the instruments much more reliable. Hiac Royco was still a major supplier of particle counters, and a second firm named Met One was also situated to provide particle counters for water treatment applications.

The interest in online particle counting was boosted further when Georgia began strongly encouraging many of its plants to install particle counters. Plants in California and a few other states also began to investigate particle counting quite seriously. A third manufacturer, Particle Measuring Systems, also known as PMS, which was well established in particle counting in the pharmaceutical industry, began offering an online particle counting system. At about the same time, the Hach Company, long the leader in drinking water turbidimetry and a major supplier of many other laboratory and process instruments, began selling a portable grab sampler manufactured by Hiac Royco. For a brief period of time, Great Lakes Instruments marketed particle counters made by Met One.

In 1994, Chemtrac Systems, a small firm known primarily for streaming current instrumentation, introduced a line of particle counting equipment, making them the first established drinking water instrumentation company to do so. In early 1996, the Hach Company began marketing an online particle counting system manufactured by PMS. As part of this agreement, PMS ceased any direct marketing in the water treatment industry.

In the biggest "off-field" development in the particle counting industry, around the end of 1995, Pacific Scientific Instruments, the parent company of Hiac Royco, purchased Met One. The ever-vigilant U.S. Department of Justice forced Pacific Scientific Instruments to sell off the Hiac Royco water treatment particle counting product line to another manufacturer, to prevent a "monopoly" in the drinking water industry. This despite the presence of Hach and Chemtrac and somehow in ignorance of the virtually complete dominance that Hiac Royco and Met One maintained in the even more lucrative hydraulics market. Somewhat reminiscent of Br'er Rabbit's admonition to "throw me in the brier patch," this move allowed Pacific to unload the increasingly uncompetitive Hiac Royco water line while remaining untouched in an industry of much greater import.

The Justice Department mandate required that the Hiac Royco line be placed with another firm with the assurance that it would maintained as a viable entity in the market. Inter Basic Resources (IBR) successfully bid for it, and has continued to market it in the drinking water industry. IBR has provided equipment related to particle counting for several years, and was a logical choice for taking over from Hiac Royco.

Just when things seemed to be settling down, two more changes have altered the particle counting landscape. The Hach Company was purchased by the parent company of Pacific Scientific Instruments, and will end its arrangement with PMS. Hach will now market the main Met One product line exclusively in the drinking water market, with Met One still active with a few of the products, as well as in other markets.

A new particle counting company emerged in 1998, Art Instruments, Inc. (ARTI), of Grants Pass, Oregon. ARTI has developed a full line of particle counting equipment for air and liquid applications. It is currently marketing drinking water systems through US Filter in North America.

At the time of writing, the current manufacturers offering particle counting equipment in the drinking water industry are Met One, Chemtrac, Hach, IBR, Art Instruments, and ATI. ATI is a drinking water instrument manufacturer that is just beginning to offer a single-channel particle counter. As is typical of "niche" markets such as particle counting, there is a good bit of "cross-pollination" between the particle counting firms, as employees change employers and bring the technology with them. Although several firms are now involved in providing particle counters, almost all of the expertise has been developed at Hiac Royco, Met One, and PMS. It is interesting that only one of these three is still even indirectly involved in the drinking water industry.

CHAPTER 15

Specifications

Table 15.1 provides a side-by-side comparison of the published specifications for the primary instruments discussed in Part III.

A. MET ONE PCX

The Met One PCX will be marketed exclusively by Hach in North America as the Hach 2200 PCX. It will be referred to by the Met One model name in this book, since it has already been on the market for a number of years. The other Met One models will still be sold by Met One. See Figure 15.1.

B. CHEMTRAC PC2400D

The Chemtrac PC2400D is the online instrument offered by Chemtrac. See Figure 15.2.

C. ART INSTRUMENTS

Art Instruments manufactures the WPC 1000 and WPC 2000. These instruments are identical except for sensitivity and flow rate. See Figure 15.3.

D. IBR WPCS

The IBR WPCS was originally designed and sold by Hiac Royco. The WPCS-01 has a smaller flow cell designed for higher concentrations than the WPCS-11.

Table 15.1 Specifications of Reviewed Instruments

Specification	ARTI WPC 2000	ARTI WPC 1000	Chemtrac PC2400D	Met One PCX	IBR WPCS-01	IBR WPCS-11
Sensitivity	2µm	1µm	2µm	2µm	2µm	2µm
Signal/noise	—	—	3:1 @ 2µm	—	—	—
Resolution	a	b	—	—	10%	10%
Coincidence	< 10% @ 15,000/ml	< 10% @ 25,000/ml	< 10% @ 15,000/ml	—	< 10% @ 18,000/ml	< 10% @ 12,000/ml
Sizing range	2 to 100 µm	1 to 25 µm	2 to 70 µm	2 to 50 µm	2 to 125 µm	2 to 40 µm
Sample flow range	50 to 110 ml/min	45 to 55 ml/min	50 to 120 ml/min	100 ml/min nominal	25 ml/min nominal	60 ml/min nominal
Flow cell dimensions	800 µm × 800 µm	600 µm × 600 µm	1 µm × 1 µm	750 µm × 750 µm	125 µm × 1000 µm	400 µm × 1000 µm
Volumetric/ in situ	Volumetric	Volumetric	Volumetric	Volumetric	Volumetric	Volumetric

a Counting efficiency: 50% at 2 µm(30 to 70% window) 100% at 5µm in the 2 µm threshold (80 to 120% window).

b Counting efficiency: 50% at 1 µm(30 to 70% window) 100% at 2 µm in the 1 µm threshold (80 to 120% window).

Note: Counting efficiency is an alternative way to look at resolution and, by implication, signal/noise. Ideally, half of the particles should fall on either side of the threshold for a given size, and none of the particles in the next highest size range. If this is achieved at the sensitivity of the sensor, it indicates a sufficiently high signal-to-noise ratio.

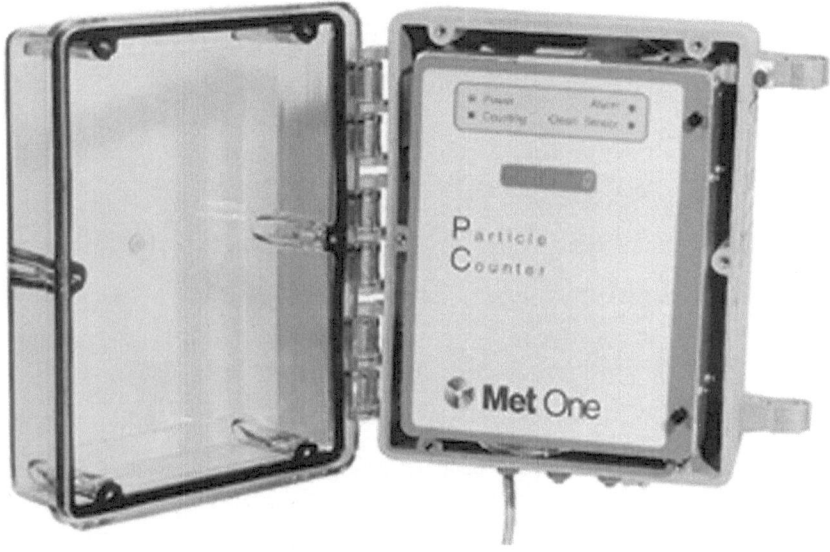

Figure 15.1 Met One PCX particle counter. (Courtesy of Pacific Scientific Instruments, Grants Pass, OR.)

Figure 15.2 Chemtrac PC2400D particle counter. (Courtesy of Chemtrac Systems, Inc., Norcross, GA.)

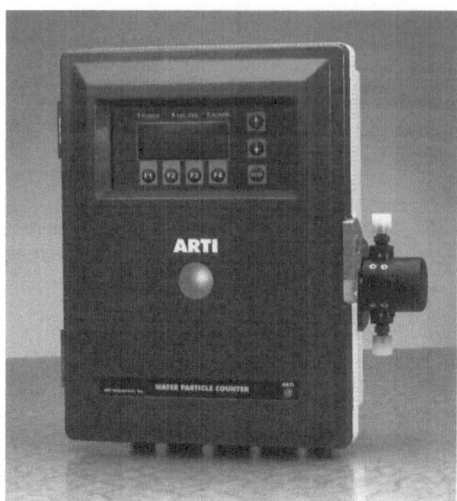

Figure 15.3 ART Instruments particle counter. (Courtesy of ART Instruments, Grants Pass, OR.)

CHAPTER 16

Particle Sensor Construction

All the manufacturers of drinking water particle counters provide particle sensors quite similar in operation and construction. The main variation to be found is in the size of the flow cell cross-section, and in the mounting of the sensor with regard to sample taps and access for cleaning. The former will mainly determine the sensitivity resolution and coincidence characteristics, while the latter impacts maintenance and potential for damage due to leaks.

A. FLOW CELL

Most of the manufacturers except for Met One provide a standard-type flow cell with synthetic sapphire windows and a stainless steel or aluminum flow cell. Met One has a sensor of this construction as well, but it is no longer supplied for drinking water applications unless specifically requested.

1. Met One

The Met One drinking water particle sensors employ a replaceable quartz cell mounted in a plastic mechanical assembly. The quartz flow cell must be replaceable because the quartz is more susceptible to scratching than synthetic sapphire. There is no advantage to this type of design outside of lower cost of manufacture, which is important in that it helps to lower the cost of particle counting for the end user.

This quartz/plastic flow cell does not have the same structural integrity of the metal/sapphire units, and is restricted to lower amounts of sample pressure. In this case, no more than 100 PSI is to be applied, and for less than 1 min. In most cases, this is many times in excess of what is required. The standard constant-head overflow weir provides an open system that limits the pressure to a few PSI. Closed systems should not be used because of the difficulty in maintaining constant pressure, and could potentially exceed the pressure limits, especially if severe water hammering were to occur. Pressurized washing to clear clogs could exceed 100 PSI, so care should be exercised. It is probably better to remove the flow cell in these cases, to avoid ruining the whole assembly.

Replacing the flow cell does leave the door open to potential problems. Care must be taken to follow the instructions carefully. If fingerprints are left on the flow cell, the optical integrity will be altered, affecting the sizing accuracy of the sensor. If the mechanical alignment is altered or loosened, the calibration will be affected, and leaks could occur. The directions for replacing the flow cell are not complicated, but if the person performing the task is not aware of the sensitivity of the instrument, carelessness could lead to problems. The sensor should be recalibrated when the flow cell has been replaced. See Figure 16.1.

The flow path for the Met One sensors is 750 μm × 750 μm. This is the same size as the stainless steel/sapphire sensors provided with earlier Met One particle counters and still available as an option. All Met One sensors are volumetric.

2. Chemtrac

The Chemtrac particle counters employ a nituff-coated aluminum flow cell. The nituff coating provides a tougher, wear-proof finish than standard black anodizing. The flow path dimensions are 1 mm × 1 mm (1 mm = 1000 μm) and the sensors are volumetric.

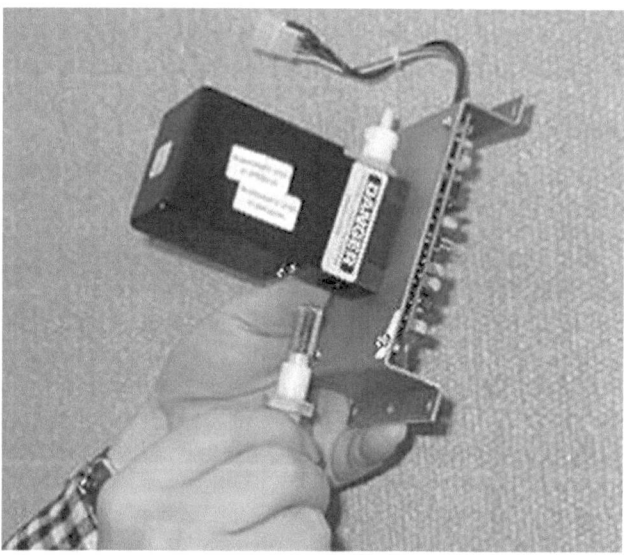

Figure 16.1 Met One removable flow cell. (Courtesy of Pacific Scientific Instruments, Grants Pass, OR.)

3. ARTI

ARTI offers two sensors for drinking water application. Each employs a standard aluminum flow cell housing with a monolithic fused silica flow cell with an antireflective coating. The WPC 2000 is designed for 2 μm sensitivity, and has an 800 μm × 800 μm flow path. The WPC 1000, a 1 μm sensor, has a 600 μm × 600 μm flow path.

4. IBR

The IBR sensor is available in two concentrations. Higher concentration limits are obtained by shrinking the flow path dimensions from 400 μm × 400 μm to 150 μm × 1000 μm.

B. CELL WINDOWS

The Chemtrac and IBR sensors come with synthetic sapphire windows, as well as the older Met One sensors. The Met One drinking water sensor employs a quartz flow cell as described above. This flow cell is a single bored piece of quartz, which is field replaceable. ARTI uses fused silica as opposed to synthetic sapphire.

C. SAMPLE FITTINGS

Sample fittings vary among the different manufacturers, with the simplest being the nylon barb fittings found on the Chemtrac particle counters. Since the Chemtrac flow connections are located outside the NEMA enclosure, the type of fitting employed is less critical, because leaks will not cause any damage or electrical hazards. The Chemtrac sensor uses standard NPT threads for mounting these fittings, so they can be easily replaced or changed to a different type if desired.

Met One also uses barb fittings on the sensor, which is located inside the NEMA enclosure. Short lengths of tubing are then run from the sensor inlet and outlet to nylon quick-disconnect fittings mounted on the walls of the NEMA box. The external sample tubing is then connected to these fittings.

IBR uses threaded stainless steel fittings designed for higher pressures. This sensor was designed for other applications requiring pressures higher than those found in drinking water plants. The same fittings are used on the older Met One sensors designed for the same purpose.

ARTI employs nylon twist-lock fittings for smooth internal flow passages designed to prevent particle buildup.

All of the sample fittings are designed for 1/4 inch outside diameter (or 6 mm metric) tubing.

D. LASER/OPTICAL ASSEMBLY

All of the manufacturers employ a similar method of mounting the laser/optical assembly to the flow cell. Mechanical configurations vary, as well as the method of alignment during calibration. This assembly is not user serviceable, and the design features are not made public.

CHAPTER 17

Particle Counter Electronics

The particle counters manufactured for drinking water all employ laser driver and detector circuits that are similar in purpose. The details of these circuits are not made available to the public, and any significant discussion of them would be beyond the scope of this book. The counting electronics and power supplies provided will be discussed briefly.

A. COUNTING ELECTRONICS

1. Met One

The Met One PCX employs a 12-bit PHA counting circuit. Up to 4096 count channels are theoretically possible. Eight channels are standard, while up to 32 are available for online use. These size ranges are set through the digital interface with the manufacturers' software. For best sizing accuracy, the upper size range should be set at or below 20 μm. This allows all 4096 channels to cover the 0 to 1 V range. If a larger upper value is used, these channels are spread out over 10 V (up to 750 μm). This reduces the sizing accuracy by an order of magnitude.

For most applications, the 20 μm upper limit is sufficient. This mode should certainly be used if resolution and count correlation between instruments is important.

The PHA card is used to calibrate the Met One sensor. All 4096 channels may be downloaded to provide a histogram of particle sizes. This allows each particle counter to be calibrated in the field.

The Met One model PCT analog output particle counter, along with the older model 215W and model 203 units, employ voltage comparator counting circuits. Each has only two size channels, which provide only total counts. Unlike the PCX, these channels are cumulative only, except for the version of the PCT that does not have a local display. That unit provides a jumper setting that allows the first channel to be set to a differential mode. The size thresholds on all of these units are set manually with trimpots. Although they may be adjusted by the user, they are not designed to be adjusted routinely. The factory should be consulted before changes are attempted.

2. Chemtrac

The 2400D counting electronics are built around a 16-bit PHA card. Only eight size channels are supported by the software, but 64,000 are theoretically possible. Once initialized, the 2400D is designed to run independently of any external control, and provides sample updates once a minute. Calibration information is stored in memory in the sensor. Size ranges are set with software.

The 2400D was initially intended to size particles up to the 1000 μm, but the circuitry to provide this was not implemented. To maintain better resolution of the smaller particles, the effective size range is from 2 to around 70 μm. This is more than enough for drinking water treatment applications.

The 2400D counting electronics is not used for calibration purposes. It is currently limited to a maximum of 16 size channels in the version configured for the grab sampler.

3. ARTI and IBR

ARTI and IBR particle counters both employ comparator-type counting electronics. Each provides eight size channels. The size thresholds are set via the digital interface.

B. POWER SUPPLY

All of the units use the newer switching power supplies that provide higher efficiency and will operate over a wide input range. Met One and IBR mount the AC supplies separately from the main enclosure to prevent hazards from sample leakage. ARTI and Chemtrac keep the sample path isolated from the main enclosure, allowing the power supplies to be mounted integrally.

CHAPTER 18

Auxiliary Features

Each of the manufacturers provides auxiliary features that will vary in importance depending on the specifics of the application. It is important to look at the way in which each of these systems performs the tasks outlined below when evaluating them. As always, consult the manufacturers directly for any updated or added features before specifying a system.

A. DIAGNOSTIC SIGNALS, ALARMS, AND DISPLAYS

All of the digital serial output systems covered provide many alarms and diagnostic signals that are displayed on the turnkey software packages provided with them. These are covered in detail in Chapter 20. This section will describe the information displayed locally on the instruments.

1. Chemtrac

The model 2400D provides all indications via the local display. Counts from each size channel are displayed in sequence, as well as the cell condition, flow rate (if an electronic flowmeter is used), and serial data address. The counts displayed are in particles per milliliter, and correspond exactly to the count values sent out to the data collection software. These counts are adjusted to the measured flow when an electronic flowmeter is used.

2. Met One

The model PCX and PCT counters provide panel indicators for power, cell condition, and count alarm as a standard feature. They are labeled differently for units not configured with the local numeric display, which might be confusing if both types of counters are used on the same system. These display lights are also used for verifying communications during system configuration.

A local numeric count display is available on both the PCX and PCT as an option. The PCX local numeric count display provides a single user-selectable size range with counts in either total or normalized values, although normalized values are determined from a fixed-flow setting, and will not respond to measured flow from an electronic flowmeter. The displayed value may not match the value sent out the serial port and displayed on the system software. The PCT display will only display cumulative total counts. Only a single size range is displayed.

The older model 215W provides no local indication of any type.

3. IBR

The IBR WPCS provides a local display that can be set to show the counts for any one of the size ranges, or the total counts. It can also be set to display the flow rate from the internal flowmeter, sample run time, alarm conditions (high flow, low flow, cell condition), alarm settings, and sensor flow parameters (for calibrating the internal flow sensor). Only one item may be displayed at a time, and is selected using the membrane keypad adjacent to the display.

4. ARTI

The ARTI WPC 1000 and WPC 2000 counters indicate counts, alarm conditions, power, and calibration status on a 4 line by 20 character display. Membrane arrow and function keys are provided to allow the user to set the display as desired.

B. SAMPLE FLOW REGULATION

1. Constant-Head Overflow Weir

All of the manufacturers provide constant-head flow regulators as standard equipment.

a. Chemtrac and Met One

The Chemtrac and Met One constant-head devices are virtually identical in design. They are adjusted by raising or lowering the sample outlet along the length of the vertical PVC pipe that makes up the body of the weir. The horizontal tube at the top of the Chemtrac weir is made of clear plastic, to allow visual indication of the amount of overflow present. Both of these devices allow the sample outlet stream to be seen by the operator, which makes quick checks of flow easy to accomplish.

These weir assemblies are designed to mount on the side of the particle counter enclosure.

b. IBR

IBR employs a similar design, with the exception that the sample outlet point is opened to the atmosphere at a fixed position. The main weir assembly must be raised or lowered to achieve the desired flow rate. The sample outlet is designed to fit into a drain assembly that obscures the flow from view. This could be corrected with a few simple pieces of hardware. Since this outlet is opened to the atmosphere at a fixed height, raising and lowering the sample drain tubing should not alter the flow rate, making graduated cylinder and stopwatch measurement less prone to error.

Like the others, the IBR weir assembly is designed to be mounted on the side of the particle counter enclosure.

c. ARTI

The ARTI constant-head weir is slightly different from the others. Flow rate is adjusted by changing the height of the weir. A simple hand-tightened ring allows the height to be changed easily. Like IBR, the sample outlet is connected directly to the drain, so that the sample flow is not open to view. A sample port is provided to allow the flow to be measured using a graduated cylinder.

The overflow drain is also hidden from view. It is concentric to the head height tube.

2. Flow Measurement and Alarm

All of the flow-metering and alarm modules provided are designed to operate along with the constant-head overflow weir.

a. Chemtrac

Chemtrac offers a low-flow alarm as a standard feature on all its counter units. This device uses a liquid level-sensing circuit along with an adjustable reservoir. Once the overflow weir has been adjusted to the desired flow rate, the overflow reservoir is adjusted until the sample covers the sensing probes. When the flow drops, the reservoir level will drop, initiating an alarm.

Chemtrac also provides an input for a 0 to 5 V flowmeter signal, which is designed to translate directly to milliliters per minute. (1 V = 100 ml/min, 1.50 V = 150 ml/min, etc.). The flowmeter input is standard on the 2400D counter.

b. Met One

Met One has a low-flow alarm, which can be added as an option. It operates in a manner similar to the Chemtrac version, except that a pressure switch is used in lieu of a conductivity sensor. It provides a contact closure that can be run to the analog input/output (I/O) board or a SCADA input.

The model PCX and 215W particle counters both provide inputs for a flowmeter signal. The PCX requires that the optional I/O card be installed to accept any input, adding additional cost unless other analog I/O are being used in the system. The Tritech flowmeter was initially developed to work with the Met One particle counters, and requires no modification.

c. IBR

Only IBR includes a built-in flowmeter as a standard item. This is actually a pressure transducer, which is accurate to only around 5 or 10%. Analog inputs are provided on the unit so that a more accurate meter can be installed. The standard flowmeter signal is used to set alarms for flow problems.

d. ARTI

An electronic flowmeter can be used with the ARTI sensor, as one of the analog inputs can be configured to accept the signal. The analog input comes as a standard feature and has the same input voltage range as the Chemtrac unit.

C. ANALOG INPUTS

All of the manufacturers provide analog inputs either optionally or as a standard part of the system. Differences involve the number of inputs provided, whether the signal is isolated or not, and whether the inputs accept current or voltage inputs.

As would be expected, only the models designed for serial data communications provide analog inputs.

1. Chemtrac

The 2400D provides up to four isolated analog inputs for connecting auxiliary 4 to 20 mA signals. All of the I/O are located on a circuit card assembly located on the back plane of the enclosure. Isolated power is provided for these inputs. The removable signal isolation modules found on earlier units have been replaced by IC sockets, which allow chips to be replaced in the field. This less-sophisticated approach is less expensive, but may result in problems if the IC chips are not handled properly in the field. These inputs are designed to accept 5 VDC signals, and allow either current or voltage inputs. Current inputs require the addition of a resistor.

2. Met One

The Met One model PCX particle counter is equipped for an optional analog I/O card which provides eight analog inputs and eight analog outputs. Only two of the analog inputs are designed to accept 4 to 20 mA signals, while the remaining six can accept inputs of +5 or +10 VDC full scale. These ranges are selected by adding or removing jumpers on the circuit board (a jumper works like a switch, but is manually positioned between fixed terminals). Only the two 4 to 20 mA signals have isolated returns, while the six voltage inputs are returned to a common ground. No signal isolation is provided.

Earlier versions of the PCX are still in service, which do not allow for addition of the analog I/O board. The analog inputs are enabled via a terminal emulation program.

The Model 215W has two 5 VDC analog inputs. One is located on the 215W counter electronics board and is intended for use with an electronic flowmeter. The other is located in the externally mounted "J-box" (see Section G below).

3. IBR

Two analog inputs are provided as a standard feature. They are designed to accept 4 to 20 mA signals directly, and separate return lines are provided. If signal isolation is desired, optional isolators can be wired into the system. These isolators may be powered off the particle counter power supply.

4. ARTI

Four analog inputs come standard, and an additional four can be added as an option. These can be configured as 0 to 5 VDC, 0 to 10 VDC, or 4 to 20 mA.

D. DISCRETE INPUTS

Discrete inputs are provided as a way of signaling backwash valve position, low flow, or some other alarm condition that will alter the particle data. Only serial data output units are equipped with this feature.

1. Chemtrac

Chemtrac provides dedicated inputs for backwash valve position and the low-flow alarm. The backwash input is designed to accept a dry contact closure. The low-flow alarm is intended only for connection of the sensing wire, and is not a standard input. Other discrete inputs have been designed into the hardware, but are not supported by the Chemtrac software.

2. ARTI

ARTI provides two discrete inputs on each unit, which can be configured by the user.

3. Met One and IBR

These units do not provide discrete inputs. Discrete signals are tied into the analog inputs using a relay contact pulled up to the appropriate voltage.

E. ANALOG OUTPUTS

Met One offers separate particle counter models for serial data and 4 to 20 mA output. The serial data model can provide 4 to 20 mA output data as well. Chemtrac, ARTI, and IBR currently offer only one model apiece, each providing both types of output.

1. Met One

a. Model PCT Analog Output Units

The Model PCT particle counters provide only two count outputs. The outputs of the PCT with local display are configured with a terminal emulation program via the RS-485 port. The full-scale (20 mA) value can be set to any value up to 9,999,999. A sample period and hold time may be programmed as well. This is used to set the update rate of the 4 to 20 mA signal.

The PCT without local display is configured with jumpers, and can be set to count for 0.1, 1, 5, or 10 minutes. The full-scale output (20 mA) can be set for 250, 2500, 25,000, or 300,000 for the first channel, and 250, 2500, or 30,000 for the second channel. The combination of settings allows for count ranges up to 30,000/ml for the first channel, and 3000/ml for the second. The jumpers may also be set to provide 4 or 20 mA test signals to allow the receiver to be scaled properly.

b. Model PCX Serial Output Unit

The optional analog I/O card provides eight analog outputs. Analog outputs may be powered from the PCX power supply, or an external power supply may be added to provide loop power. Each output may be set to a specific particle size range (differential or cumulative), and full-scale value (count level corresponding to 20 mA). These counts are not normalized for the flow rate, but represent total counts. These output signals are all returned through the common ground point of the instrument.

The particle count values from the analog outputs are calculated independently of those from the serial output. They will not match the values displayed on the data collection software or the local display. The outputs can be placed in a test mode to provide either 4 or 20 mA signals to allow scaling of the receiver. This is done using a computer terminal emulation program connected to the RS-232 port of the PCX.

2. Chemtrac Model 2400D Serial Output Unit

The 2400D provides up to four self-powered analog outputs, which correspond to the first four size ranges programmed into the particle counter. The 4 to 20 mA output span is set to 200, 2000, or 20,000 counts/ml by plugging resistors into the circuit card. This rather crude arrangement will result in additional signal error, since the lower count ranges are scaled down from the full 20,000 range by a simple voltage divider circuit.

3. IBR

The IBR WPCS counter can be configured for two or four analog outputs. An optional analog output module must be mounted in the enclosure and wired into place. The analog outputs correspond to the first two or four size channels configured into the counter. The output is scaled logarithmically according to the formula:

$$\text{Loop current in mA} = 4 \times [(\log_{10}(\text{counts} + 10)]$$

The output ranges from zero up to almost 100,000 counts. However, a full-scale output of 20 mA is used to represent an alarm condition, such as a cell condition error, overconcentration, sensor failure, or flow out of range by more than 25%.

When the counter is run in "Dataloop" mode, the analog output range may be set to any desired upper and lower count range. This mode is used when the particle counters are not running with the IBR supplied software.

4. ARTI

ARTI has an optional analog I/O card that provides four analog outputs and four analog inputs.

F. DISCRETE OUTPUTS

Discrete outputs are designed to turn on alarms or otherwise signal events of some sort. These are found on analog output units, to provide indication of instrument problems or alarm conditions. Serial output units provide alarm information as part of the serial data output.

1. Met One

The Model PCT with local display provides a single discrete output for signaling alarm conditions. All alarms are tied together to this single point. It is an open collector solid-state relay that is built into the circuit board.

The other Met One particle counters do not provide discrete outputs.

2. ARTI

ARTI provides two discrete outputs for control.

3. IBR and Chemtrac

IBR and Chemtrac do not provide discrete outputs.

G. ENCLOSURES AND PACKAGING

Packaging is integral to functionality. All the units are designed for use in filter galleries and other damp environments. All of them are housed in NEMA-rated enclosures, designed to protect the electronics from humidity and sprayed water.

It is an advantage to be able to clean the flow cell without opening the NEMA enclosure, since water may be spilled into the enclosure. Chemtrac and ARTI offer the easiest access to the sensor, which sticks out of the NEMA enclosure. Clogs or partial obstruction may require opening the NEMA enclosure of these other units. Met One and IBR have pressure limitations that could be exceeded if compressed air is used to clear a clogged flow cell.

1. Chemtrac Systems

Since sample does not pass through the NEMA enclosure, all the AC power and RS-485 communications lines can be located in a single box. Conduit fittings are mounted on a removable panel located on the bottom of the NEMA enclosure. This panel can be taken off, leaving all the conduit in place. The RS-485 communications and I/O signal lines are connected to keyed removable connectors, which can be removed from the circuit board. The same is true of the discrete signals. Only the AC power wires must be removed individually.

2. ARTI

The WPC 1000 and WPC 2000 also provide an external sensor to avoid sample flow through the NEMA enclosure.

3. Met One

In all the Met One particle counters, the particle sensors are located inside the NEMA 4X enclosure, allowing for the possibility of sample leakage into the enclosure. For this reason, the AC power supply is located in a separate enclosure. This supply comes standard in a non-NEMA-rated package, but a NEMA-rated power supply is available as well. The NEMA supply is recommended for use in the filter gallery, or in any other location where the possibility of coming in contact with water exists. This supply should always be mounted above the flood level of the gallery. A 6-foot power cable is supplied, and the user may install a longer cable if required.

All external signals and power are wired to screw terminals in the main enclosure. They must be completely removed and reinstalled when the unit is taken out of service.

An optional RS-485 "J-box" (junction box) is available for the PCX, allowing it to be removed from the serial data highway without breaking the signal line. This option should be employed whenever possible, as it greatly simplifies service. Extra J-boxes can be installed in additional locations around the plant to allow counters to be moved for temporary sampling.

The 215W requires a J-box for connection to the RS-485 network. The power supply is located in this enclosure.

4. IBR

AC power is located in a separate NEMA enclosure, since the liquid sample passes through the primary NEMA enclosure. The sensor is located along the bottom of the NEMA box, allowing access to the sample inlet port without opening the enclosure. The counting electronics are mounted in the upper half of the same enclosure. A desiccant is provided to absorb humidity, and provides visible indication of when it needs replacement through the front panel of the enclosure.

All external signals and power are wired to screw terminals in the main enclosure. They must be completely removed and re-installed when the unit is taken out of service.

CHAPTER 19

Serial Communications

All of the manufacturers provide units designed for digital serial communications. The great majority of applications will be found to require the speed and efficiency of serial data transfer. For this reason, each of the manufacturers provides serial data communications as an integral part of its base system. Each of the standard turnkey systems incorporate RS-485 multidrop communications designed for direct computer interface. This fact alone should be enough to convince those still wed to 4 to 20 mA interfaces of the inefficiencies of such an approach. If it were cheaper or easier to follow that path, at least some of the manufacturers would be doing so.

A. INTERFACE

All of the units provide a standard RS-485 interface. This allows for multiple units to be connected to a single-twisted shielded-pair cable. The number of counters that can be connected to the system depends on the number of instrument addresses available, and the limitations of the data collection software. All of the manufacturers allow for up to 32 units. More units may be added to each system, but will have to be placed on a separate RS-485 network.

ARTI, Chemtrac, Met One, and IBR provide RS-232 access to the particle counter. This allows individual parameters to be set without connecting to the RS-485 network. Only one unit may be connected at a time on an RS-232 network. Chemtrac provides a separate connector, while Met One and IBR require moving a couple of jumpers to switch from RS-485 to RS-232.

RS-485 communications can be conducted over two- or four-wire cabling. Two-wire communications involves transmitting and receiving data over the same pair of wires, while the four-wire arrangement keeps transmission and receiving lines separate. There is no difference in speed, as the units do not transmit and receive at the same time. ARTI, Met One, and Chemtrac are designed to operate on a two-wire arrangement, although the Chemtrac unit may also be operated in four-wire mode by removing jumpers on the printed circuit board. IBR requires a four-wire connection. The only difference in these arrangements is the slightly higher cost of four-conductor cabling.

B. PROTOCOLS

Each of the available units communicates data by way of a unique protocol. When a standard turnkey system is installed, the protocol is not important from the user's standpoint. Only when the particle counters are to be tied into SCADA or a data collection system does the protocol come into play.

Only ARTI and Chemtrac use protocols developed to simplify interface with third-party equipment. ARTI has implemented the Modbus RTU protocol, which is supported by almost all of the available SCADA software packages. Chemtrac uses a modified version of the optomux protocol. While the standard optomux protocol is supported by many SCADA packages, not all of the commands are included. The standard optomux protocol provides for minimums, maximums, and averages to be transmitted by the field hardware, whereas the SCADA software programs perform these tasks themselves. Often these commands are left out of the driver packages that come standard with the SCADA software. Unfortunately, the particle counters make use of these commands for other data functions, and so the drivers are not fully compatible. However, since many of the commands are functional, the existing drivers can be modified to support these commands, as opposed to being written from scratch.

Certainly at some point, other units will be designed to communicate via more widely accepted protocols. Most of these will be built around popular programmable logic controller (PLC) protocols such as Modbus, since the amount of data communicated is rather large, and fits the format of the PLC.

The individual protocols are too complex to provide within this book. Met One includes detailed information on protocols in their standard equipment manuals. The other manufacturers should be able to provide this information on request.

CHAPTER 20

Manufacturer's Software

All of the major manufacturers provide a software data collection package as part of their turnkey system. For the most part, these packages perform the same tasks in logging and displaying the particle count and related data, and printing reports. The software packages differ in the way in which they structure the data, as well as in operational aspects.

Software is often the deciding factor when purchasing a turnkey system. This is understandable, as the software is the most visible part of the system, and provides the daily point of contact for the operator. This is also true because so little is understood about the particle counting equipment. It is our hope that this book will help to alleviate this problem. At any rate, it is important to look closely at the particle counting software, with consideration for the ease of use and access to the information, as well as the usefulness of the many features involved. A plethora of bells and whistles is not necessarily an indication of a more-advanced product. It may be less important to have a fancy system than one that performs the basics and is easy for the operators to learn and work with. Other applications may require all the advanced features available.

We are not attempting to rate the packages described below, and particular comments about the usefulness of certain features should be taken as opinions, which may not be applicable to every situation. Our intent is to aid the reader in evaluating the relative importance of features, many which may not otherwise become apparent until months after the system is in operation. These comments should only be taken as a starting point. Most of the manufacturers provide demo versions of their software programs, and will gladly provide them to potential customers. Take advantage of this and become familiar with the way these packages work. Let other operators try them as well, to get a broader perspective. The computer expert in the plant may really like an intricate system, which will never be utilized properly by the rest of the operations staff.

We also do not undertake to rate how effectively each of the features described perform in day to day operation. Software is constantly being upgraded, and is never "bug-free." Problems encountered at the writing of this book may have been fixed by the time the book is published.

A. OVERVIEW OF AVAILABLE SOFTWARE

1. Met One

Met One currently produces a package that they call Water Quality Software (WQS). It is available in four different configurations, named Insight, Vista, VistaNet, and VistaNet Server. These are basically four different grades of the same package, with Insight designed for small systems of seven particle counters or fewer. Insight has fewer features (no backwash indication, for example). Vista supports up to 32 particle counters, where VistaNet is a networked version designed for systems having from 20 to 100 units. VistaNet Server will accommodate up to 200 particle counters per server. It is designed for large multiplant water systems.

For simplicity, these packages will be called Met One WQS in the descriptions below. They must be operated under the Windows 95, 98, or NT operating system. Some older 16-bit versions are still in use, but should not be specified for new installations. WQS was written in-house, and has an attractive and well-written manual.

2. Chemtrac

Chemtrac provides a package known as TracWare, which was developed specifically for the Chemtrac particle counters by Instrumentation Design, Inc., of Atlanta, GA. It was written with design input from the engineering staff that developed the particle counters, and is licensed exclusively to Chemtrac. For all intents and purposes it can be considered an "in-house" package. It can support up to 32 particle counters. TracWare is a 16-bit program written for Windows 3.1 and 3.11, and will run under Windows 95 and 98 as well. It can be provided in a networkable version. It is intuitive and easy to use, which is fortunate, as the manual is functional but poorly reproduced.

3. ARTI

The newest program of the four reviewed is the ARTI Aquarius software. It supports up to 32 particle counters. Aquarius is a 32-bit program designed for use with Windows 98 and NT. It includes extensive context-sensitive help files and an adequate manual. A unique feature is its ability to be run in different languages. Currently, English and Japanese are fully implemented, and the code can adapt to other languages. The program will run under the language set up as the Windows default, or can be set during program launch with a single-command line entry. A simple language translation utility is used for customizing the program for other language and character sets.

4. IBR

IBR supplies Intellitest software. Intellitest is created from a generic software program named LabView™, which is a product of National Instruments. LabView is widely used for data collection, and can be configured to provide a highly customized package. The Intellitest program provides a compact display of the particle count data, and is designed to operate with a commercial spreadsheet program to provide more-advanced reports and data display. The recommended spreadsheet is Microsoft Excel™, although other brands may be employed. Intellitest is a 16-bit program compatible with Windows 3.11, 95, 98, and NT. No networked system is yet available, but special programming is available to allow the package to share data with some SCADA programs. The program is easy to use and fairly intuitive, which is good since the manual provides only minimal information. Users are on their own with the spreadsheet program.

Some of older IBR systems use the AccuCount software developed by Hiac Royco out of a third-party SCADA package called Fix DMACS. It is less functional and more difficult to configure than the other packages. SCADA packages are designed to be used for a wide range of processes and applications, so must be more open ended to allow for this. It usually requires several steps to perform a simple task with one of these packages. The raw SCADA package is configured by the particle counter manufacturer to perform the basic particle counting tasks. If end users want to change some things about the system, they may have to learn how to configure the SCADA package to do what they want. This approach is expensive and cumbersome, and has been largely abandoned in the industry. For this reason, IBR developed the Intellitest software. AccuCount is a 16-bit Windows 3.1 and 3.11 package, which should be capable of running under Windows 95.

B. FEATURES

All the packages run under some version of Windows operating system, which allows for multiple programs to be operated simultaneously. However, all of them recommend that the data collection computer be used for just that, and that other programs be run on other machines where possible. This is a wise course to follow, especially with the price of computer hardware being so low. Larger systems should use networked machines where possible, so that reporting and data analysis can be performed on a machine different from the one used for data collection, to minimize problems.

All of the software programs provide trend plots, tables, status or current value screens, and event logs. Alarms are displayed in various forms, and reports are provided. The major features of each package are presented below, with notable features highlighted.

1. Data Presentation

All the packages are operated out of a fixed reference window, which provides access to all the features. All use the standard Windows-type drop-down menus across the top of the display. WQS and Aquarius provide icon-driven selections as well, while TracWare and Intellitest use labeled buttons that are kept in the same location for each screen. Consistency is important for ease of use, while the exact layout and style may be more a matter of user preference.

Two of the packages provide for multiple displays to be viewed simultaneously. These are the Met One WQS and ARTI Aquarius. See Figure 20.1. The other packages provide only one screen at a time, although several forms of data may be presented on that screen. Up to eight screens can be displayed in Aquarius. The WQS software can provide up to four simultaneous graphs or data tables, depending on the size and resolution of the monitor used. Multiple displays allow for direct comparison of different data points, as well as comparison of historical data with current data for the same sample points.

Aquarius utilizes the standard Windows MDI (Multiple Document Interface), so that the user can shrink or enlarge any of the data windows using the mouse. Screens may also be moved around and rearranged as desired. Tool bar icons allow screens to be tiled or cascaded with a single mouse click. Met One limits graphs and data screens to 3/4 or full-screen size only. Multiple graphs are shrunk vertically to fit the display, but the horizontal size remains fixed. The relative positions of the displays are also fixed, with the current data screen on the left-hand side of the display, and the graphs or data tables on the right. See Figure 20.2.

IBR Intellitest utilizes a fixed screen with an upper and lower half. The upper half remains fixed, and contains most of the buttons for selecting the various options and displays. The lower half changes to provide those displays. Buttons are dimmed and disabled when they do not apply to the current display, a feature that helps simplify operation.

2. Trend Display

As mentioned in previous sections, the trend display of data is the most useful and informative. All the software packages provide some form of trend display. In every case, the vertical axis is used to indicate data values, while the timescale is displayed on the horizontal axis. All of the packages provide different scales for both axes, which are user selectable.

a. Scaling and Configuration

Chemtrac's TracWare, although limited to a single trend display, provides individual data scales for each of the four pens on the trend display. See Figure 20.3. This is especially useful when trending particle counts along with an auxiliary input, such as turbidity. Finished water turbidity is usually below 0.1 NTU, whereas particle counts may be over 1000/ml or more. Each of these trends can be scaled independently to provide good resolution. The other programs use a single data scale for all pens, which can be set to linear or logarithmic. In the above example, multiple graphs would be used to achieve maximum resolution with Aquarius and Met One WQS.

The vertical data scales on the TracWare program are adjusted up and down by clicking on the top or bottom scale value for the relevant pen. The scale values can

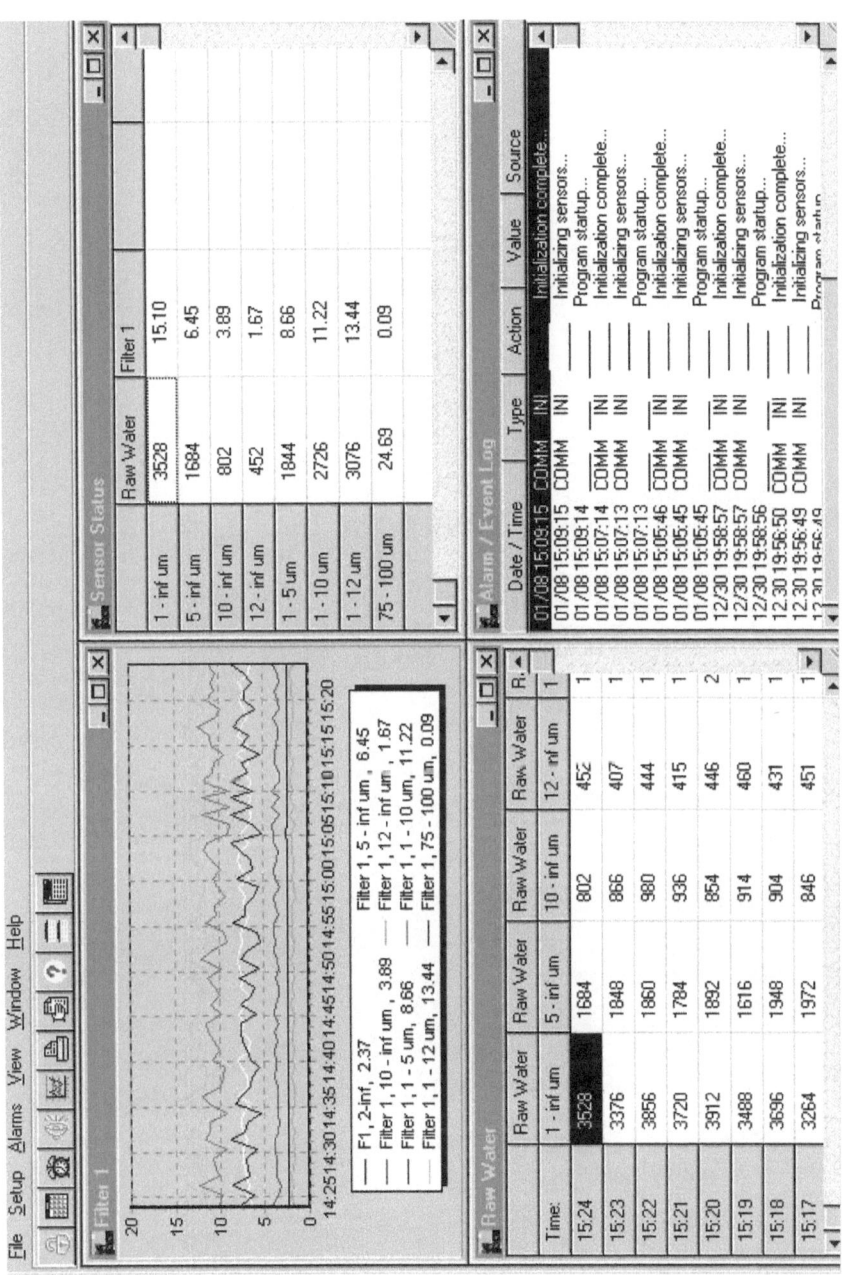

Figure 20.1 Aquarius main display. (Courtesy of ART Instruments, Inc., Grants Pass, OR.)

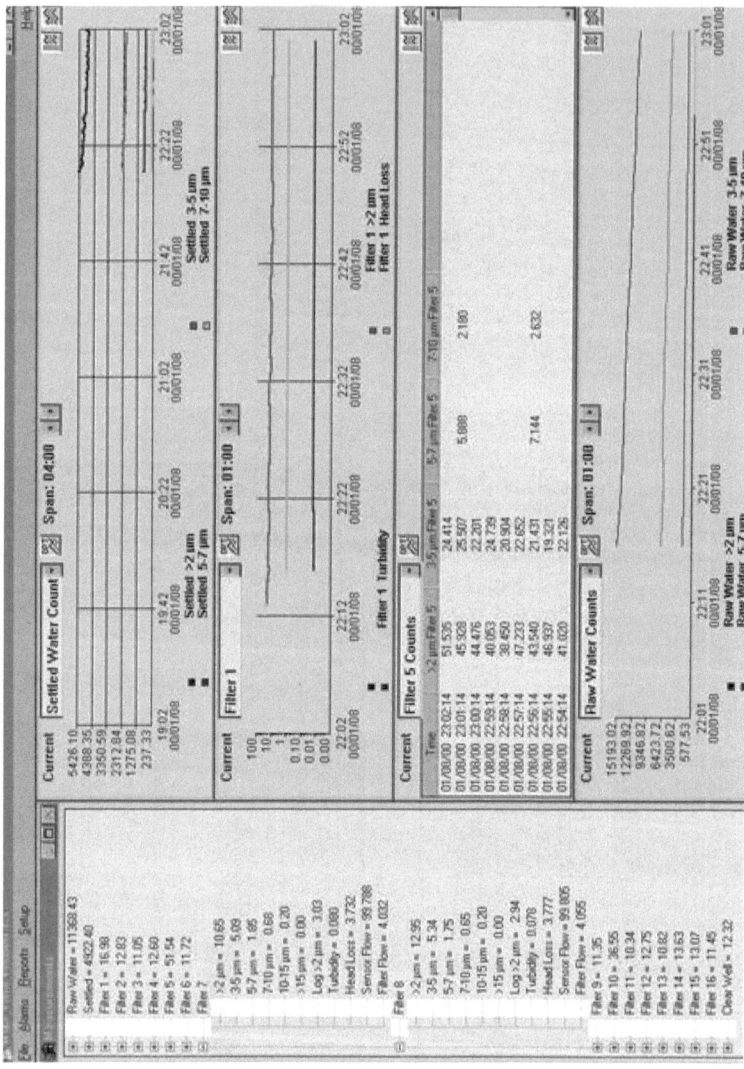

Figure 20.2 WQS main display. (Courtesy of Pacific Scientific Instruments, Grants Pass, OR.)

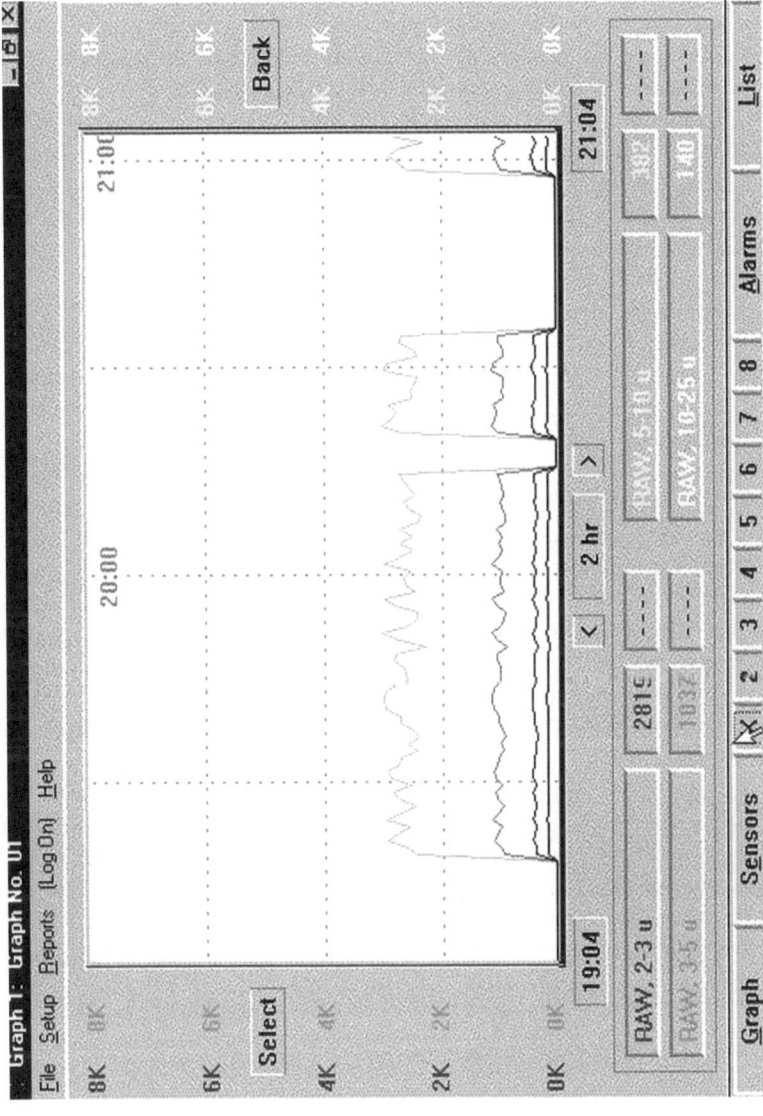

Figure 20.3 TracWare trend display. (Courtesy of Chemtrac Systems, Inc., Norcross, GA.)

also be set on the graph setup menu, along with a choice of colors for each pen. This latter feature allows for consistent color codes to be used on multiple graphs when the same parameter is displayed more than once. With the exception of TracWare and Aquarius, the other programs use set colors for each pen, which requires that the same value be matched to the same pen on each graph if consistency is desired. Intellitest provides entry fields for the upper and lower scale ranges, which can be accessed directly on the trend display. A configuration menu for Intellitest, Met One WQS, and Aquarius allows for choosing between log and linear scaling. Met One and IBR provide an automatic scaling option, which adjusts the scale to the highest value displayed. Aquarius and Intellitest feature "zoom" options, which let the user enlarge a part of the graph for detailed viewing. Intellitest has a "grabbing" option, which allows any point on the trend line to be moved to the desired part of the display, automatically adjusting the axes for optimal viewing. Aquarius allows the data in the "zoomed" window to be moved around as desired. See Figure 20.4.

Intellitest allows for display of only one particle counter at a time. The analog inputs connected to that counter may be trended along with the particle count data, which may be displayed as differential or cumulative counts versus time. A histogram provides counts versus size. To view comparative trends, a commercial spreadsheet must be utilized.

The other packages allow for trending data points from anywhere in the system together on the same display. TracWare and WQS provide four pens per display, Aquarius and IBR Intellitest, eight. All the packages use individual colored pens. All but TracWare provide individual line markers for black-and-white printing of the trend data.

The pen selection is similar for the Aquarius, TracWare, and WQS programs. The latter two provide for up to 32 trend displays with any combination of data inputs or calculated (i.e., log removal) values assignable to any trend graph. There is no practical limit in the Aquarius software. Each trend graph can be given a name for easy reference. Each sample point is given a tag name when the system is initially configured, and these tags are selected from a pull-down menu when setting up individual graphs. For particle count values, a separate menu is provided with the particle size ranges. All of the graphs configured are then available from a pull-down menu, making selection simple. The only significant difference between these packages is that TracWare provides a set of eight buttons at the bottom of the trend display which allows direct access to the first eight graphs configured. There are also up and down arrows that allow the user to scroll through all 32 graphs in sequence. See Figures 20.5 and 20.6.

b. Time Period

The amount of data collected by each system varies, with some programs providing user settings to control the sample period. This, in turn, impacts the time period that can be displayed on a single trend display, as well as the resolution of the data. TracWare collects data every minute, and there is no user option for selecting a longer or shorter sample period. Data are displayed on the trend graph in increments of 1, 5, 10, 15, 30, and 60 minutes. This value is chosen by clicking

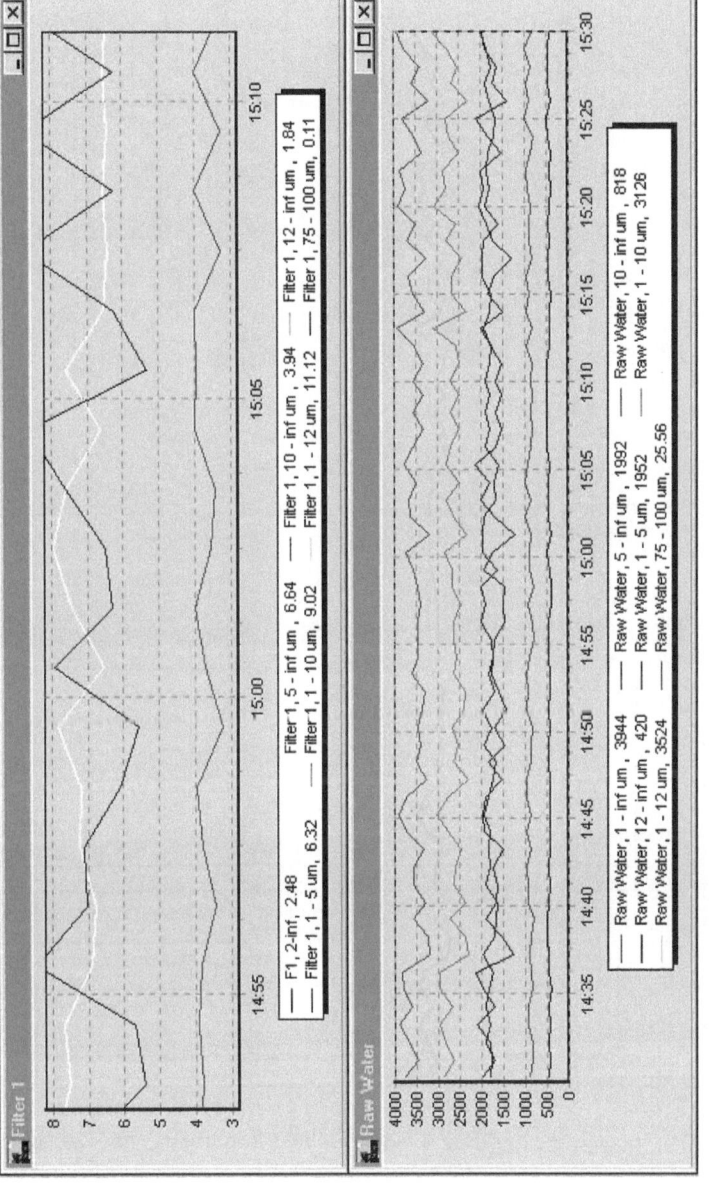

Figure 20.4 Aquarius zoom feature. (Courtesy of ART Instruments, Inc., Grants Pass, OR.)

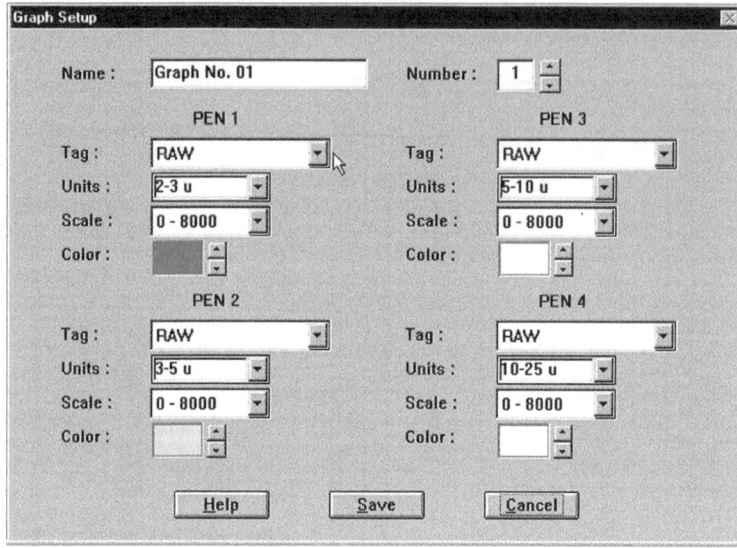

Figure 20.5 TracWare trend graph setup screen. (Courtesy of Chemtrac Systems, Inc., Norcross, GA.)

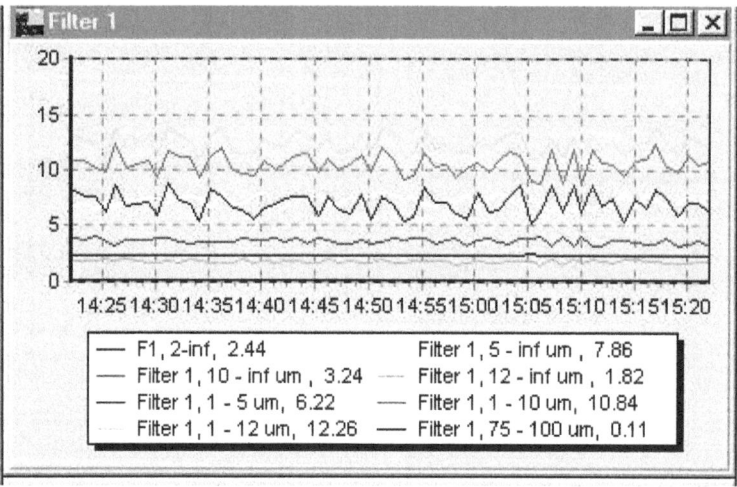

Figure 20.6 Aquarius trend display. (Courtesy of ART Instruments, Inc., Grants Pass, OR.)

on right and left arrows beneath the trend display. In all, 120 sample points can be displayed on the screen, allowing for time spans from 2 to 120 hours (5 days). The current value is always displayed on the far right side of the display, and the display shifts to the left when a new sample is recorded. A separate historical program has been added that allows the user to set the time period desired to view the data stored

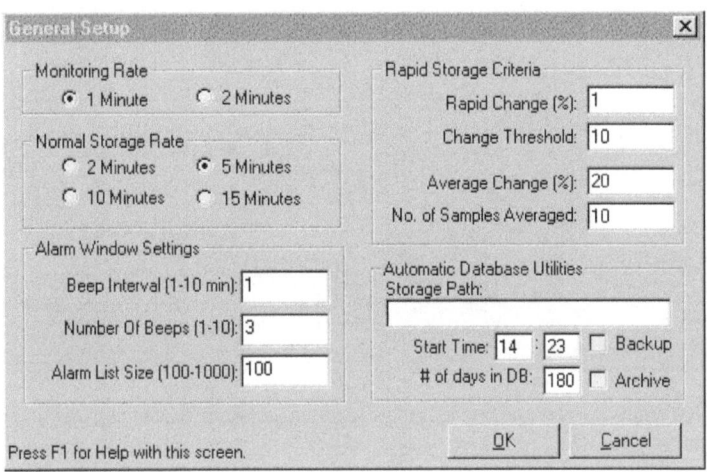

Figure 20.7 WQS rapid storage configuration. (Courtesy of Pacific Scientific Instruments, Grants Pass, OR.)

during that period. This historical program runs separately from the main TracWare program, but uses similar data display functions.

Met One WQS has a unique feature known as "Rapid Storage." It allows for an increase in the amount of data stored based on user-defined parameters. For example, if the particle counts exceed a certain amount or change by a given percentage, the software will store data at an increased rate to provide better resolution of the data. Under normal conditions, fewer data are collected, for the sake of economy. In this system, the data are monitored every 1 or 2 minutes, as selected by the user. It is stored every 2, 5, 10 or 15 minutes on the 1-minute cycle, or at twice those values when set to monitor data every 2 minutes. When the preset alarm limits are exceeded, the data are then stored at the monitoring rate (every minute or 2 minutes). See Figure 20.7. The WQS graphs can hold up to 2000 data points per measurement, and can display up to 22.7 days worth of data. The length of time that can be displayed will depend on the sampling frequency.

Three trend display modes are provided. When set to display updated values, the time frame can be adjusted using right and left arrows at the top of the display, but each click shifts the data by only 1 min. Clicking directly on the span value brings up an entry field where a longer span can be input. A second mode provides a scroll bar, which allows rapid scanning of the last 2000 samples stored. The current data are not updated in this mode. The third mode is used to retrieve long-term historical data. Beginning and ending sample times and dates are input into entry fields. Met One combines historical and real-time data in the same package, as each display can be configured in one of the three modes. Several other options are available, such as manual or automatic scaling for the data (vertical) axis. Vertical and horizontal grids can be added, and a graph title and label for the vertical axis can be added for printouts. Trends can be displayed in two or three dimensions, and the traces in thin or thick line widths. See Figure 20.8.

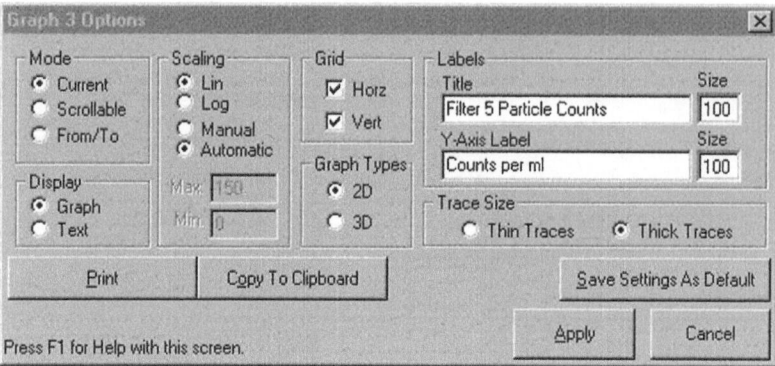

Figure 20.8 WQS trend graph configuration. (Courtesy of Pacific Scientific Instruments, Grants Pass, OR.)

Aquarius displays up to 240 sample points ranging from 1 hour up to 5 days. For time displays larger than 4 hours, any of four different data presentations are available. For example, a display of 24 hours would mean that each displayed sample point is spaced 6 minutes apart. The display can be set to show the actual count values taken every 6th minute, or the minimum, maximum, or average of the samples collected during that 6-minute period. In fact, each of the four can be displayed in separate windows for the same data point, allowing for direct comparison of the four options. This feature is especially valuable for looking at time periods spanning several days.

3. Tabular Data Display

Tabular data are useful for viewing actual values in relation to trend graphs. A certain event or unexpected change in data can be viewed in tabular form to get a more exacting representation of the data. The data value should be presented along with the sample time to allow for pinpoint measurement of the time of the occurrence.

Aquarius, WQS, and TracWare provide tabular data in a method closely tied to their trend-graphing functions. The Aquarius trend graph can be directly changed to tabular data by double-clicking the display (and visa versa). WQS provides a simple option on the graph configuration menu to allow the display to be switched between trend and tabular data. The Rapid Storage works the same as for the trend display, so some columns may have more data than others. If the Rapid Storage is engaged, the data will be displayed at 1- or 2-minute intervals for the affected data point, while the other points will show data only at the normal data collection rate. See Figure 20.9.

TracWare provides for 32 tabular displays, configured in the same manner as the trend graphs, with the exception of pen selections. Eight data points can be placed on each table, and can be set to display data in increments of 1, 5, 10, 15, 30, and 60 min. Up to 240 values can be displayed, allowing for time spans ranging from 4 hours to 10 days. The latest value appears at the top, with older values descending downward.

File Alarms Reports Setup			
Current	**Filter 1**		
Time	>2 µm:Filter 1	Turbidity:Filter 1	Head Loss:Filter 1
01/08/00 22:49:14	18.765		
01/08/00 22:48:14	15.206		
01/08/00 22:47:14	17.914		
01/08/00 22:46:14	18.565	0.081	2.237
01/08/00 22:45:14	18.667		
01/08/00 22:44:14	15.383		
01/08/00 22:43:14	20.594		
01/08/00 22:42:14	18.737		
01/08/00 22:41:14	21.742	0.085	2.185
01/08/00 22:40:14	20.896		
01/08/00 22:39:14	20.370		
01/08/00 22:38:14	17.260		
01/08/00 22:37:14	18.215		
01/08/00 22:36:14	19.709	0.079	2.255
01/08/00 22:35:14	20.721		
01/08/00 22:34:14	20.290		
01/08/00 22:33:14	17.714		
01/08/00 22:32:14	22.952		
01/08/00 22:31:14	24.317	0.084	2.258
01/08/00 22:30:14	22.075		
01/08/00 22:29:14	20.162		
01/08/00 22:28:14	22.398		
01/08/00 22:27:14	20.595		
01/08/00 22:26:14	20.117	0.084	2.200
01/08/00 22:25:14	22.374		

Figure 20.9 WQS tabular data display. (Courtesy of Pacific Scientific Instruments, Grants Pass, OR.)

A vertical scroll bar is used to view values extending below the viewing area. Just like the graph display, eight buttons at the bottom of the screen allow for direct access to the first eight data tables, and arrows allow for scrolling between all those configured. These eight buttons are not tied to the same data points as those used for the graphs, unless configured in that manner by the operator. See Figure 20.10.

Aquarius, WQS, and TracWare provide easy methods for comparing data between sample points. The long vertical tables with eight values side-by-side in Aquarius and TracWare provide an easy way to compare multiple data points over a long time span, while the multiple displays in WQS allow for quick comparisons, although only four values are presented side-by-side, and the gaps from the Rapid Storage may make time comparisons a little less direct. These systems allow for displays to be configured quickly, so that points of interest can be placed on the same display with a few strokes of the mouse. Figures 20.11 and 20.12

IBR Intellitest only provides tables for current values. To view historical tabular data, the spreadsheet option must be used. The spreadsheet can be activated from the main menu of the program, providing integrated operation. Macros are available for some common spreadsheet programs.

4. Status Display and Alarms

Aquarius and TracWare provide the most complete single-status display, given that all the current values for eight particle counters are presented on a single display.

Select	RAW 2-3 u		RAW 5-10 u		RAW 25-50 u		RAW 80-100 u	
Time:		RAW 3-5 u		RAW 10-25 u		RAW 50-00 u		RAW 2-100 u
21:11	2459	905	333	122	45.0	16.6	6.10	3887
21:10	2552	939	345	127	46.8	17.2	6.33	4034
21:09	2310	850	313	115	42.3	15.6	5.73	3651
21:08	2328	856	315	116	42.6	15.7	5.77	3680
21:07	2571	946	348	128	47.1	17.3	6.37	4063
21:06	2249	828	304	112	41.2	15.2	5.58	3555
21:05	2309	849	312	115	42.3	15.6	5.72	3649
21:04	2819	1037	382	140	51.6	19.0	6.99	4456
21:03	2236	822	303	111	40.9	15.1	5.54	3534
21:02	2665	980	361	133	48.8	18.0	6.61	4212
21:01	2799	1030	379	139	51.3	18.9	6.94	4424
21:00	2628	967	356	131	48.1	17.7	6.51	4153
20:59	2489	916	337	124	45.6	16.8	6.17	3934
20:58	0.00	0.00	0.00	0.00	0.00	0.00	0.00	0.00
20:57	0.00	0.00	0.00	0.00	0.00	0.00	0.00	0.00
20:56	0.00	0.00	0.00	0.00	0.00	0.00	0.00	0.00
20:55	0.00	0.00	0.00	0.00	0.00	0.00	0.00	0.00
20:54	0.00	0.00	0.00	0.00	0.00	0.00	0.00	0.00
20:53	0.00	0.00	0.00	0.00	0.00	0.00	0.00	0.00

Figure 20.10 TracWare tabular data. (Courtesy of Chemtrac Systems, Inc., Norcross, GA.)

Time:	Raw Water 1 - inf um	Raw Water 5 - inf um	Raw Water 10 - inf um	Raw Water 12 - inf um	R.
15:25	3784	1672	990	484	2
15:24	3528	1684	802	452	1
15:23	3376	1848	866	407	1
15:22	3856	1860	980	444	1
15:21	3720	1784	936	415	1
15:20	3912	1892	854	446	2
15:19	3488	1616	914	460	1
15:18	3696	1948	904	431	1

Figure 20.11 Aquarius tabular display. (Courtesy of ART Instruments, Inc., Grants Pass, OR.)

This includes all size ranges and analog input values, along with communications status, discrete inputs, and cell condition. TracWare displays sample flow in milliliters per minute if an electronic flowmeter is used, or a simple alarm if the standard low-flow alarm is employed. Cell condition is displayed on a 0 to 100% scale, to keep things simple. If all four analog inputs are in use, the status display will be too large for the screen, and will have to be scrolled down to allow the analog values to be seen. If more than eight sensors are installed on the system, a second page of eight is reached by clicking on an arrow at the bottom of the display. Up to four pages (32 particle counters) are available. See Figure 20.13.

Current	Settled Water Count ▼ 🖼			
Time	>2 µm:Settled	3-5 µm:Settled	5-7 µm:Settled	7-10 µm:Settled
01/13/00 23:13:24	4775.550		720.200	228.175
01/13/00 23:11:26				
01/13/00 21:56:17				284.750
01/13/00 21:55:17	5622.525	2417.850		279.425
01/13/00 21:54:17	5467.025	2508.850	816.250	285.675
01/13/00 21:53:17		2473.000		284.775
01/13/00 21:52:17		2391.475		279.975
01/13/00 21:51:17	5659.650		853.475	
01/13/00 21:49:18				
01/08/00 22:45:14	4984.875		717.275	
01/08/00 22:44:14		2162.650	760.225	248.525
01/08/00 22:43:14		2222.750	743.075	240.225
01/08/00 22:42:14	4917.275	2153.675	760.100	252.225

Current	Filter 5 Counts	▼ 🖼		
Time	>2 µm:Filter 5	3-5 µm:Filter 5	5-7 µm:Filter 5	7-10 µm:Filter 5
01/13/00 23:13:24	40.805	13.485		
01/13/00 23:11:27				
01/13/00 21:56:17	72.966	35.305	9.797	
01/13/00 21:55:17	84.870	39.953	10.502	
01/13/00 21:54:17	90.486	32.927	11.602	4.761
01/13/00 21:53:17	85.713	35.689		
01/13/00 21:52:17		38.939	14.358	
01/13/00 21:51:17	88.327	29.300	12.507	
01/13/00 21:49:19				
01/08/00 22:45:14		24.942		
01/08/00 22:44:14	60.917	19.654		
01/08/00 22:43:14	47.975	26.920		
01/08/00 22:42:14	55.175	18.893	8.670	

Current	Filt 1-4 Turbidity	▼ 🖼		
Time	Turbidity:Filter 1	Turbidity:Filter 2	Turbidity:Filter 3	Turbidity:Filter 4
01/13/00 23:11:27				
01/13/00 21:55:17	0.086			
01/13/00 21:54:17	0.076	0.078	0.083	0.084
01/13/00 21:53:17	0.080			
01/13/00 21:52:17	0.079			
01/13/00 21:51:17	0.083			
01/13/00 21:50:17	0.083			
01/13/00 21:49:19				
01/08/00 22:41:14	0.085	0.085	0.085	0.083
01/08/00 22:36:14	0.079	0.075	0.083	0.095
01/08/00 22:31:14	0.084	0.079	0.083	0.083
01/08/00 22:26:14	0.084	0.076	0.091	0.078
01/08/00 22:21:14	0.081	0.078	0.078	0.083

Figure 20.12 WQS tabular data display. (Courtesy of Pacific Scientific Instruments, Grants Pass, OR.)

In TracWare, the particle counter tag names run across the top of the status display, and change colors with the status of each unit. Black indicates an operating unit, red indicates a unit which is off-line (not communicating with the software), and yellow indicates a filter in backwash mode. Alarms are indicated across the bottom of the display.

The Met One WQS status display is located along the left-hand side of the main display. It is covered only when the trend graphs are extended to full-screen mode. The data are arranged according to particle counter, from top to bottom, in the order that the particle counters are initially configured in the system. Clicking on the tag name for a given sensor will reveal or hide all but the first size channel (usually total counts) of the current value data for that sensor, including all size ranges and any analog inputs connected to that unit. Depending on the number of counters, as well as the size ranges specified and the number of analog inputs used, only a fraction of the status data can be displayed at once. It is easy to run through the various particle counters to check the current values. See Figure 20.14.

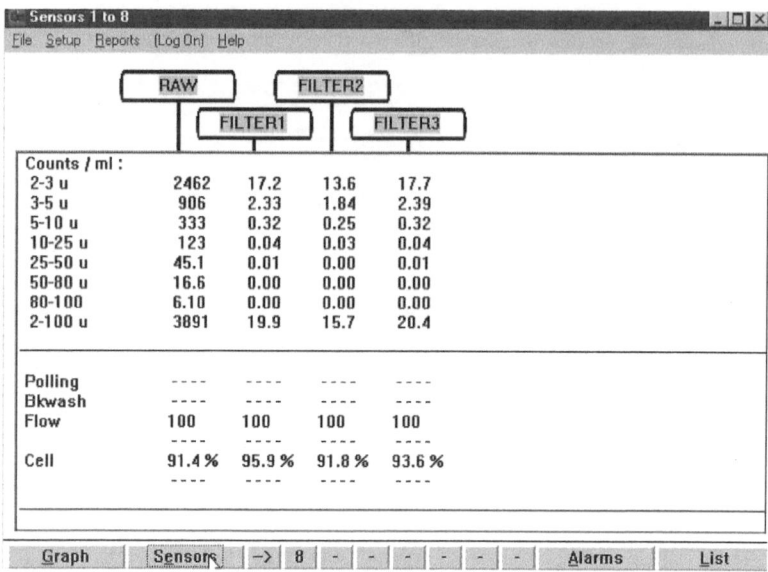

Figure 20.13 TracWare sensor status screen. (Courtesy of Chemtrac Systems, Inc., Norcross, GA.)

Aquarius, WQS, and TracWare provide similar alarms, for communications problems, cell condition, and high and low values for analog inputs. WQS provides these for particle counts and log/percent removals as well, while TracWare provides only high alarms for particle counts, and a low alarm for log/percent removals. Aquarius allows two alarms per variable, which can be set to any combination of highs and lows, including high, high, or low, low. Low particle counts are usually a function of sample flow or equipment problems, which are covered by other alarms. High removal rates would be the same. One nice feature employed in Aquarius and TracWare is an alarm deadband range, which is used to keep an alarm from constantly restarting after it is acknowledged. The deadband provides for an additional number of counts or percentage of change before alarming again. See Figure 20.15. WQS does not provide a numerical cell condition.

WQS displays alarms in a pop-up window. WQS provides several options for audible alarms. If the computer is configured with a sound card, recorded sounds can be played to signal the alarms. If no sound card is available, a series of time-delayed beeps can be set. TracWare will provide a beep alarm until an acknowledge is received.

Intellitest displays the current value for the upstream particle counter (usually raw or settled water sample) on the upper half of the main screen. Up to four downstream units are displayed across the lower half, with the next group of four displayed by clicking a button on the lower left-hand side of the display. One of the eight particle size ranges is displayed for each particle counter, along with the sample flow rate, and the two analog input values. Clicking the View Expand button on any

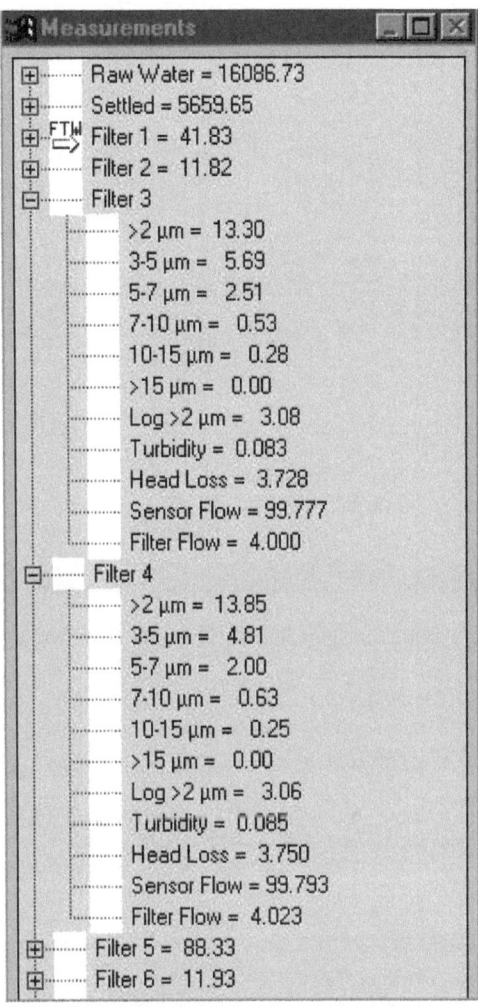

Figure 20.14 WQS sensor status display. (Courtesy of Pacific Scientific Instruments, Grants Pass, OR.)

of these counter displays produces a pop up display with current values and log removals for up to eight size channels.

Five alarms are indicated along the right-hand side of the upper window. One is for high particle counts, two for analog inputs, one for sample flow, and the last is used to indicate communication errors. If one of these alarms is signaled, the indicator turns from green to red. An indicator will also turn from green to red on the display for the affected particle counter. The left-hand indicator is for communication alarm, and the right-hand indicator denotes a data alarm. The combination of the main alarm display and the individual alarm indicators provides the exact location and type of alarm for each event.

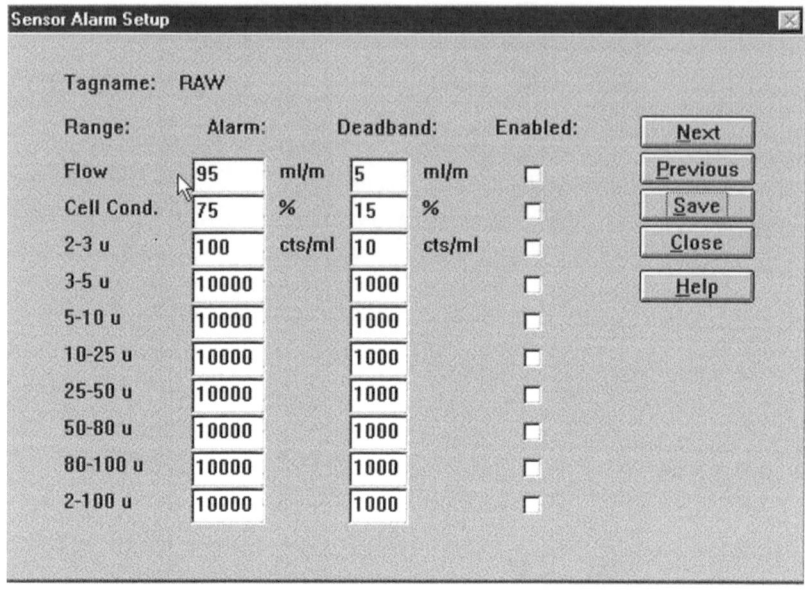

Figure 20.15 TracWare alarm configuration. (Courtesy of Chemtrac Systems, Inc., Norcross, GA.)

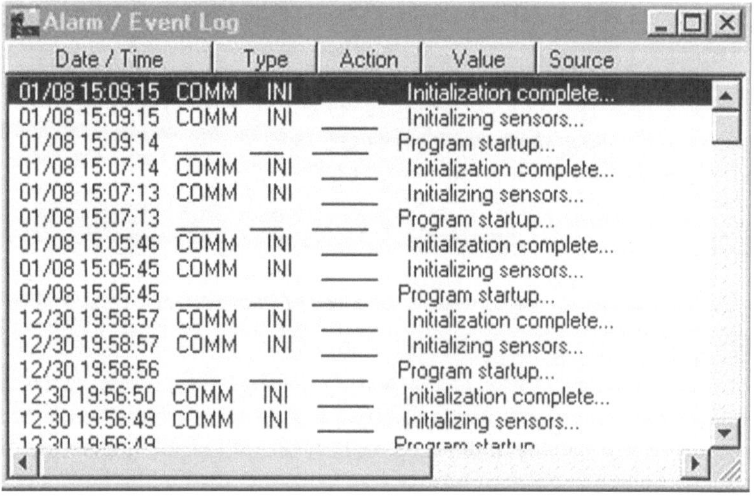

Figure 20.16 Aquarius alarm/event log. (Courtesy of ART Instruments, Inc., Grants Pass, OR.)

5. Event Log

The Aquarius, WQS, and TracWare provide event logs for recording events and alarms. TracWare allows the operator to input messages into the log, to record various events or descriptions that might be useful when the data are analyzed later. See Figure 20.16.

WQS displays the last 100 to 1000 (set by the user) alarm events on the alarm pop-up window. In all three cases, the event/alarm log is stored on the hard disk for future reference.

6. Reporting

No drinking water system would be complete without reporting functionality. Part of the difficulty in designing a good reporting system is that any number of formats may be employed, and there is no established reporting method to provide useful guidelines. The most current packages have added reporting features, which are configured in a manner consistent with the rest to the software, to keep the user from having to learn an additional spreadsheet or other type program to generate reports. Those who prefer to use a spreadsheet will find that data can be exported from all these packages.

a. WQS and TracWare

Met One WQS and TracWare provide the most similar reporting features, as might be expected. These programs both incorporate a utility called Crystal Reports, which is designed to supplement custom programs with preconfigured reporting features. Crystal Reports provides an "engine," which takes selected data and converts it into report pages that can be customized for any number of styles of presentation. It provides a standard viewing feature, which allows the user to page through the report, zoom in and out, and otherwise examine the report on screen before printing it out. It likewise provides conversion of the data into any of a number of user-selectable formats compatible with the major spreadsheet and word-processor programs, as well as generic formats (comma separated, tab separated, etc.)

Crystal Reports and similar engines allow the particle counting software manufacturers the benefit of providing advanced functionality without having to "reinvent the wheel." All that is required of them is to write an interface to the rest of the software, which provides the desired data to the reporting engine, and to configure the report styles to fit the application. Both WQS and TracWare have written interfaces that function along the lines of the other parts of the package.

WQS provides a report screen which looks and functions like the status display on the main operating screen. See Figure 20.17. Report groups that consist of any number of report formats and sample measurement points are set up by the user. This information is then listed down the left side of the report configuration. Different levels of detail are revealed or hidden by clicking on the various group names. Under each group name is a listing of the reports to be produced, and a listing of the

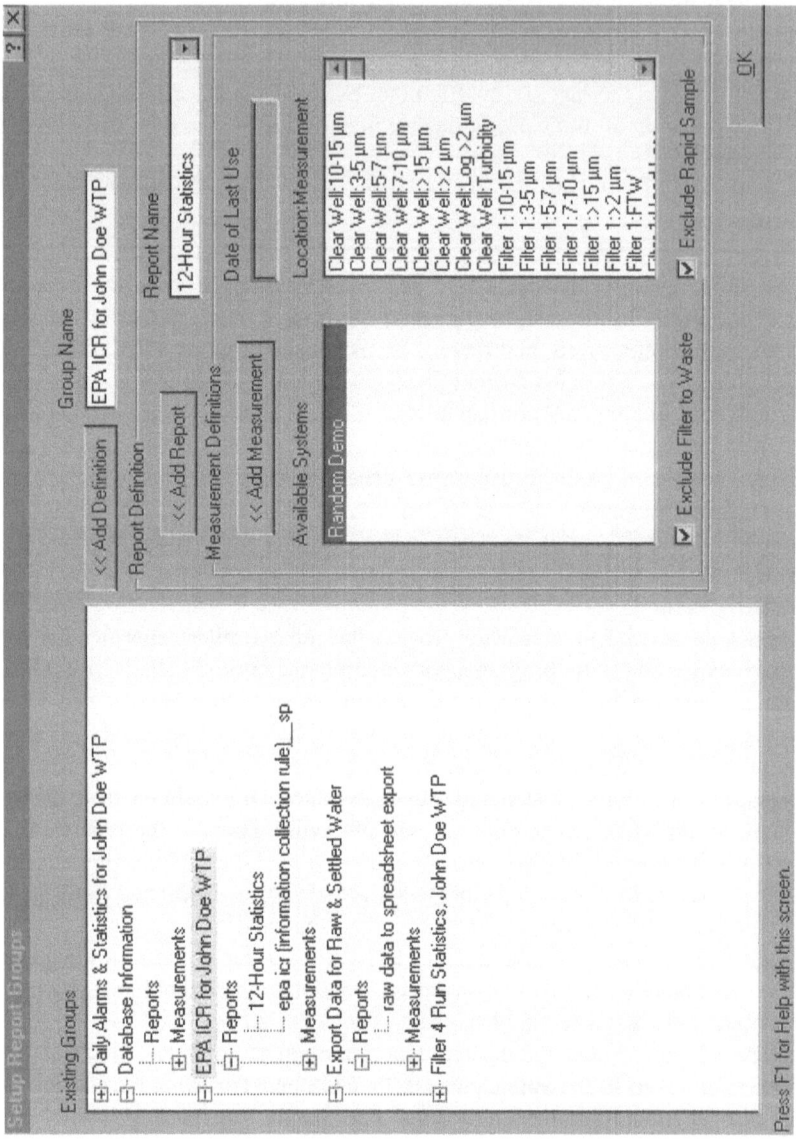

Figure 20.17 WQS report configuration screen. (Courtesy of Pacific Scientific Instruments, Grants Pass, OR.)

measurement points to be included in the report. The reports themselves are configured for various lengths of time, such as hourly statistics, 2-hour statistics, 24-hour
statistics, etc. Alarm reports and customized reports designed for the ICR (Information Collection Rule) are available as well. Each set of measurements can be set to
include or exclude filter-to-waste data (if the particle counters have been connected
to the backwash valves), as well as the Rapid Sample data described above. The
printout will indicate which of these options has been selected, but there is no way
to review the report setup on screen. Each of these report groups can be scheduled
to print at a certain time of day, or viewed, exported, and/or printed on demand. The
created groups all appear on the scheduled print list, initially at time 0000. This time
must be changed to activate the report creation. Midnight is not allowed for report
creation, because of the many system overhead operations that are carried out to
begin each new day. If several reports are created each day, the times should be
spaced apart to avoid overloading the system. Reports should not be created during
scheduled database backups.

TracWare uses a report setup menu that is similar to that used for the tabular
data screens. See Figure 20.18. As many as eight sample points can be placed on a
single report, and both sensors and particle size ranges (where appropriate) are
selected from pull-down menus. Up to 48 "Quick Reports" can be configured. The
sample period can be selected for 1, 5, 10, 15, 30, or 60 min, just as in the trend
and tabular display windows. The date and time for the newest and oldest data to
be included are chosen by clicking on up and down arrows. Up to 144 samples can
be listed on a single report, and the number of samples for each report is determined
by the time and sample period settings selected. A report title may be entered, and
the report format selected from a pull-down menu. The Quick Reports can be printed
or exported to disk in the same manner as in the Met One WQS software.

Scheduled Reports are selected from the Quick Reports, and can be set to run
at a given time each day. These reports may be printed, saved to disk in a selected
data format, or both. The data in each daily report cover the time span set in the
Quick Report setup, except that the dates are rolled ahead to remain current.

While Chemtrac particle counters are equipped with backwash detection and
TracWare records and tags the data appropriately, there is no setting allowing that
data to be ignored by the reports. To achieve this, a special report format template
must be designed using an off-the-shelf version of Crystal Reports. Any number of
report templates may be added in to the system. Templates may also be added to
WQS, although no support is provided by Met One. Met One will provide custom
templates at an extra cost. See Figure 20.19.

TracWare provides more flexibility in selecting the amount of data displayed,
while WQS comes with more preconfigured templates. While both packages provide
plenty of options, those requiring specific report configurations should look at the
inexpensive Crystal Reports package, or convince one of the manufacturers to produce
a custom template (preferably before the order is finalized, to make sure that it can
be done within the constraints of the system, and while leverage is the greatest).

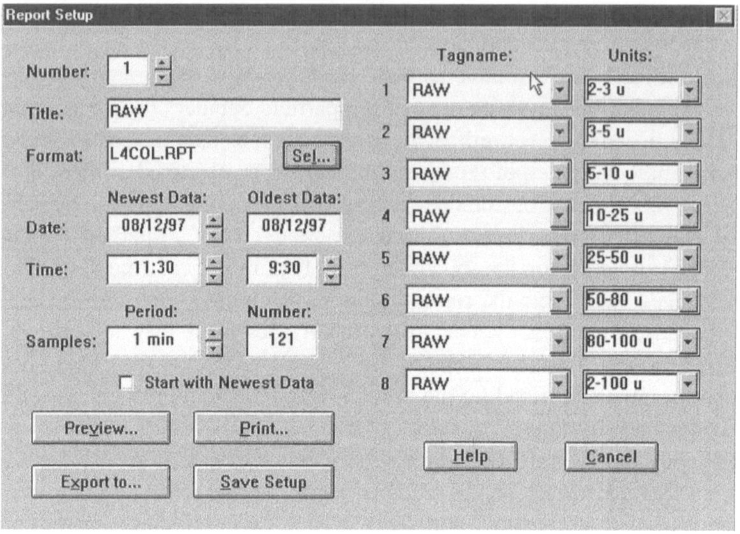

Figure 20.18 TracWare report configuration. (Courtesy of Chemtrac Systems, Inc., Norcross, GA.)

b. Intellitest

Intellitest provides a couple of built-in reports. The Daily Report is an hourly minimum, maximum, and average for all eight size ranges, the two analog inputs, and the log removals for each particle counter. The daily totals are provided at the bottom of the report. The Monthly Report follows the same format, except that daily values are displayed, with the monthly totals at the bottom. There is no automatic scheduling option. No type of preview option is available. Other types of reports must be generated with a separate spreadsheet program.

c. Aquarius

Aquarius is similar to TracWare in the way that reports are configured, except that no Quick Report templates are available. As many as 64 tags can be placed on a report, and any number of scheduled reports can be configured. Four basic data types are available for export: ASCII (CSV), Dbase IV, Foxpro, and Paradox.

C. CONFIGURING THE SOFTWARE

Initial as well as subsequent configuration is usually performed on site by the operators. All of the programs are configured in similar fashion. Configuration is not an everyday operation, so it is less critical when evaluating these systems, as long as it is not unduly complicated.

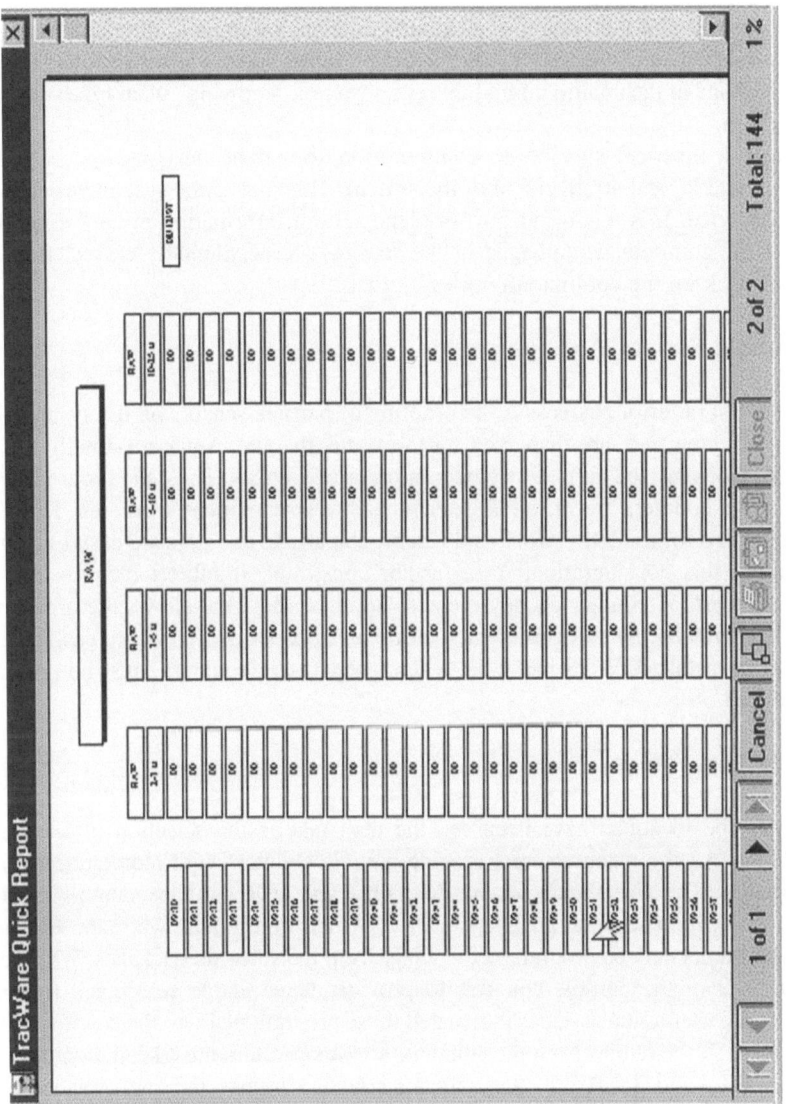

Figure 20.19 TracWare report. (Courtesy of Chemtrac Systems, Inc., Norcross, GA.)

In every case, the software identifies and communicates with the particle counters according to the address of the counter. Each of the particle counters in the system must be configured with a unique address. The tag names and other items related to each particle counter are all resident on the computer, and have no direct bearing on the operation of the system. For example, the analog input scaling and tag names are all resident in the software. The particle counter returns a value corresponding to the voltage at the analog input, but has no way of "knowing" what type of scaling is involved.

From a practical standpoint, configuration should be designed to achieve an efficient and logical arrangement of the system. This makes the system easier to use and to service. It is a good idea to save the setting information in hard-copy form in case the computer is damaged or the settings are accidentally erased. It is also wise to back up the configuration files on diskette.

1. Particle Sensor Arrangement

The first order of business is the ordering of particle sensors on the system. This will affect how they are displayed on the status display. Aquarius, Intellitest, and TracWare display the particle counters in order of address. Thus, the order in which the particle counters are installed determines the arrangement on screen. However, the addresses for the IBR, ARTI, and Chemtrac particle counters are easy to change, mitigating this consideration. As a further constraint, Intellitest requires that the upstream particle counter occupy the zero address. Met One allows the ordering to be determined by the software, which eliminates concerns over where each particle counter is installed. WQS has a list where the particle counters must be placed in the desired order.

2. Size Thresholds

Once the locations have been set, the next task is the selection of size range thresholds. All the manufacturers provide particle counters with remotely selectable size ranges. The size thresholds for Met One and IBR particle counters must be configured with a separate program, such as Windows terminal, while the Chemtrac and ARTI units may be programmed directly from the software packages. WQS reads the sizes from the particle counters to provide the available selections. Intellitest requires the user to input values that match those programmed into the counter. It does not read back the ranges from the counters, so the operators must keep them straight.

Aquarius and TracWare constrain all counters to use the same ranges. This minimizes errors, at the expense of some flexibility. The Met One and IBR systems will run with whichever ranges are programmed into them. In most cases, the same ranges will be set in all the particle counters.

3. Passwords and Security

All the major programs provide password security to limit access to the program. These are designed to keep untrained operators from accidentally changing the configuration of the program. Software passwords are not foolproof, and are not intended to provide protection from someone bent on causing problems or intent on defeating them. If such concerns exist, the computer should be physically locked up where access is controlled.

TracWare provides passwords for up to four users, and like WQS which provides only one, there is only one level of password protection, which provides access to configuration and display setup commands. Aquarius has two levels, one for changing the configuration, and a second for operator use, which allows temporary changes in scaling.

4. Additional Features

a. Print Features

All the programs will print out reports, as well as trend plots. The WQS and Aquarius packages provide higher-resolution trend printouts than does TracWare, which uses screen prints. Screen prints are lower resolution, because they print the display directly to the printer. All the programs will print color trends, which makes viewing multiple data much easier.

Aquarius provides an impressive array of options for printing trends in almost any form imaginable. A WYSIWYG print preview allows for quick verification. See Figure 20.20.

b. Histograms

IBR Intellitest provides histogram displays. Histograms provide a three-dimensional view of the data, and are useful for judging the relationship between data points at a glance. For example, all the size channels could be displayed for a given particle counter, providing a good picture of the particle size distribution.

5. Removal Calculations

The primary data calculation is based on particle removal, with log or percent removal being the two methods employed by all the programs. The setup for these calculations is almost identical in the Aquarius, WQS, and TracWare programs. An influent and an effluent particle counter are selected, along with the desired size range. WQS allows different size ranges to be selected for each of the particle counters, while Aquarius and TracWare do not. There is no reason to choose separate size ranges, and the latter approach prevents accidental selection of incompatible ranges. All three packages provide an option for log or percent removal. Log is better for graphing, while percent may be easier to understand for those unfamiliar with logarithms.

One of the yet-unresolved issues in particle counting is what to do with log removals when no particles are present downstream of the filter. This is a common occurrence, especially in the larger size ranges. TracWare and Aquarius use 4 log as a default value in this case.

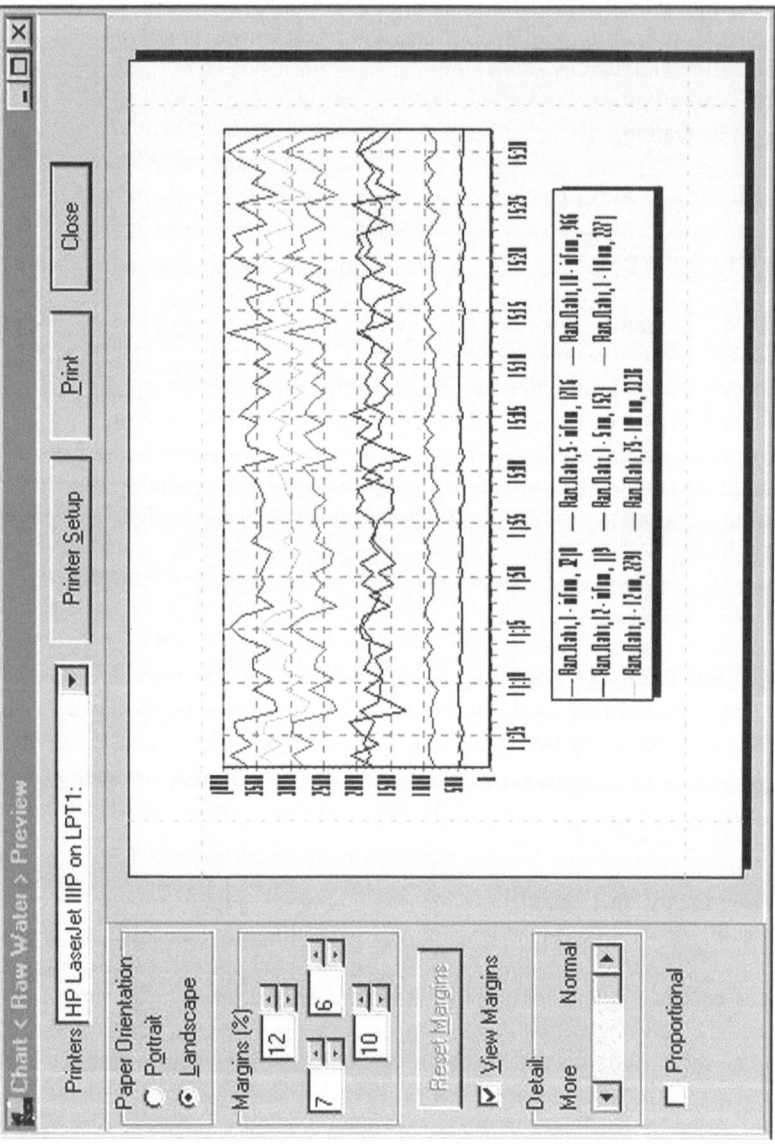

Figure 20.20 Aquarius print preview. (Courtesy of ART Instruments, Inc., Grants Pass, OR.)

CHAPTER 21

The Complete System

Now that the various features of the different brands of equipment have been presented, it is time to put them all together and take a brief look at the total system. The reader should bear in mind the limitations of such an undertaking, as features change and new models are introduced. If this book is used properly, enough knowledge should be gained to allow one to evaluate these changes sufficiently. It would be unfair to the manufacturers and the end user to take the brief comments made in this chapter as sufficient grounds for making a decision. These comments are provided to help the reader to see better the "big picture" when undertaking a thorough evaluation. They are not based on exhaustive empirical studies, or head-to-head tests on every feature. The manufacturers specification's and claims have been taken at face value, and have not been independently verified. The information presented is based on informed opinion, gathered through much experience with a broad spectrum of operators and applications, but still subject to the many biases and flaws of fallen human nature. Caveat Emptor.

A. THE TREATMENT PLANT/APPLICATION

All the equipment reviewed is designed for drinking water treatment applications. As such, each should perform adequately. Most of the important decisions will have to be based upon the service record and equipment reliability of each manufacturer, which will have to be accessed by consulting references. It is important to consult a sufficient number of references to overcome individual biases and offset the occasional "black hole" installation that will happen to everyone. Make sure that proper account is taken for the age of the equipment installed. Many improvements have been made in the past few years, as the market has matured. Often complaints stem from frustration with the functionally limited equipment sold in earlier years, especially before adequate computer interface was provided. Many operators have never properly learned how to use the equipment, and blame the equipment unfairly.

Complaints about Hiac Royco early on helped Met One to get a foothold in the market. Similar complaints are now heard about some of these early Met One applications. While Hiac Royco made little attempt to improve its products before selling the line to IBR, Met One has made great strides in the past few years. It is important to get enough information to take these factors into account. On the other hand, do not assume that the largest number of installations necessarily means the best product.

Most of the other considerations will have to do with the ability of the operators to operate and maintain the system. Some of the systems require a little more effort to clean and maintain, and a particular style of software presentation may be more suited to a given set of operators. Although it is becoming less practical to have each manufacturer set up a demonstration system on site, each will provide demo software, which should be set up for the operators to try out. Get copies of the manuals in advance. The "simpler is better" approach is not always true. Assuming that a computer will make the system too complicated is not usually wise. It is precisely the ability to see the trended particle counter data that makes the equipment easier to use. The same goes for auxiliary inputs. Trending particle counts and turbidity together will be the best way to help the operations staff assimilate the technology. Do not skimp on training. Initial and follow-up training will make a great deal of difference.

Most of the turnkey systems are designed to accommodate a large number of particle counters, so adding units piecemeal or after a plant expansion should not be a problem. If the turnkey system is to be used for only a couple of years before a complete SCADA integration is undertaken, the various SCADA integration options should be studied in detail prior to making the initial purchase. As this is usually the least-understood aspect of the technology, it is easy to "worry about it later," which is unwise. Although the manufacturers typically provide less-than-optimal support for SCADA integration, it stands to reason that the larger firms should be able to handle such requests more easily. Met One has certainly had the most opportunity to have developed expertise in this area, with the largest installed base of equipment in the water market, as well as experience in other market segments. The ARTI use of the Modbus RTU protocol is a welcome development. The Chemtrac protocol was built around the optomux standard in a conscious attempt to make SCADA integration easier, but most of the standard SCADA optomux drivers are incomplete and must be modified to accommodate it. Any of the manufacturers may have developed an interface for a particular SCADA package, so it never hurts to check.

B. EQUIPMENT FEATURES

1. Packaging

Packaging is integral to functionality. All the units are designed for use in filter galleries and other damp environments. It is an advantage to be able to clean the flow cell without opening the NEMA enclosure, since water may be spilled into the enclosure. ARTI and Chemtrac offer the easiest access to the sensor, which sticks out of the NEMA enclosure, while the others allow access to cleaning without opening the NEMA enclosure. Clogs or partial obstruction may require opening the NEMA enclosure of these other units. Met One and IBR have pressure limitations that could be exceeded if compressed air is used to clear a clogged flow cell.

The removable bottom panel on the Chemtrac counter allows analog I/O signals to be quickly disconnected and reconnected via plug-in connectors. The IBR and Met One enclosures leave the power supplies mounted and isolated, but require individual removal of the analog signals.

2. Sensor Characteristics

All of the units use light-blocking particle sensors. All of them work according to the same principles outlined in Part I of the book. The largest deviations appear in the use of less-rugged materials in the Met One sensor, and the nonvolumetric design of the older Hach/PMS equipment. The former is important as a maintenance consideration only, in that it must be treated with more care than the others. The nonvolumetric design of the older Hach sensor could result in the loss of sizing resolution accuracy. Until tighter standards are developed, this will not be an issue.

3. Counter Electronics

For most online applications, the type of counter electronics is not critical, as only a few size ranges are ever used. Chemtrac and Met One employ multichannel analyzers (MCA) type counters, as opposed to the standard comparator types. Met One uses the MCA card for calibration, which allows the complete system to be optimized for better performance.

Chemtrac, ARTI, and Met One all offer versatility with regard to analog I/O.

4. Software

A thorough evaluation of the software provided by each manufacturer shows that less difference exists than one may expect. While certain features may be more appealing to a given operator, all the manufacturers provide the important features necessary to a well-run particle counting system.

ARTI, Met One, and Chemtrac offer similar packages, with Aquarius and WQS being a little more up-to-date with multiple graphing features. The IBR Intellitest program is not as developed as the others, but is functional, and will likely keep evolving.

All in all, our recommendation would be that the software offered by all of the manufacturers is adequate, and should not be the driving factor in the system selection. All other things being equal, then perhaps the look and feel of a given program might be a deciding factor.

5. Experience

Met One is the leader in providing in-house expertise. It is the company with the most installations, as well as the longest tenure in drinking water applications. It is larger than the other companies, and should be strengthened by its merger with Hach. Hach has a tremendous installed base of turbidimeters and other plant instrumentation.

Chemtrac is a small company with a good marketing channel but with little technical expertise. If serious efforts at tightening standards are pursued by the industry, it is doubtful that Chemtrac will be able to keep up. Equipment must be improved over time, or it will become dated. The Chemtrac equipment had some excellent features when it was introduced, but it has never matured as a product. The IBR equipment has been around since 1988, and it remains to be seen to what degree it will be improved and advanced.

ARTI has a staff of experienced designers, many having been formerly associated with Met One. Its alliance with US Filter will give it a good presence in the market.

CHAPTER 22

Grab Samplers

All four of the manufacturers offer grab samplers, with Met One, Chemtrac, and ARTI leading the way. IBR still offers the Hiac Royco unit originally designed for Hach as the "Log Easy." Each of the units is similar in size and weight, and operates in like manner.

A. EQUIPMENT FEATURES

The ARTI WQA 2000, Met One WGS-267 Grab Sampler, and the Chemtrac 2400 PS are designed to operate in grab or continuous, online mode. The IBR can as well, although its internal pump cannot be bypassed. The pumps in each of these units are not designed for continual operation, so it is recommended that long-term online applications be set up using a constant-head overflow weir and gravity flow.

1. Sample Delivery System

All of the units employ small DC gear pumps that pull the sample through the sensor. The speed of the pump is controlled by a DC voltage source, which is not user adjustable. The ARTI WQA 2000 and the Met One WGS-267 employ a user-adjustable needle valve rotometer to allow the operator to adjust the flow rate. The other units provide no flow adjustment or display. The gear pumps provide a smooth, quiet, pulseless flow and can clog if large concentrations of particles are present. They are designed for around 400 hours of operating life, which will cover several years worth of grab sampling, but only a few weeks of continuous operation.

The ARTI, Chemtrac, and Met One units are easily converted to continuous, online operation by moving a couple of pieces of tubing. This allows for the pump to be bypassed, and connection made to a gravity flow system.

All of the units operate off AC power and rechargeable DC batteries. Battery operation provides ease of portability, as well as reducing the electrical shock hazard when operated in a wet environment.

2. Packaging

The Chemtrac 2400 PS is packaged in a standard NEMA 4X plastic box, with the sensor extending from the side. This allows the sample flow to be completely separated from the battery and electronics. The AC power adapter is a sealed "desktop" power supply similar to the type used for laptop computers. This smaller NEMA box is mounted on a simple adjustable stand and stored in a rugged watertight ABS plastic equipment case for ease of storage and portability. The 2400 PS has a local display and a membrane touch keypad operator interface. No printer is available. See Figure 22.1.

The Met One WGS 267 is built into a folding deep-drawn aluminum case, with the counting electronics mounted in the door, and the sensor, power supply, battery, and sample delivery system in the larger body of the case. Unlike the Chemtrac unit, the model 267 must be taken apart to access the sensor for cleaning. Liquid does pass into the case, allowing for the potential for leakage into the power supply. A fault isolation circuit breaker is built into the power cord to shut off the system if the power supply is shorted. A membrane keypad and local display are provided, along with a small thermal printer. See Figure 22.2.

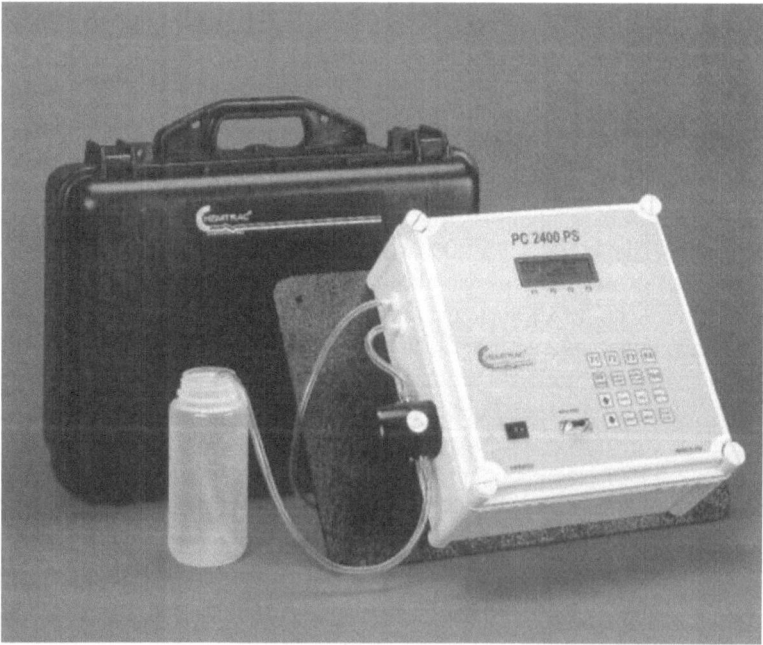

Figure 22.1 Chemtrac PC2400PS grab sampler. (Courtesy of Chemtrac Systems, Inc., Norcross, GA.)

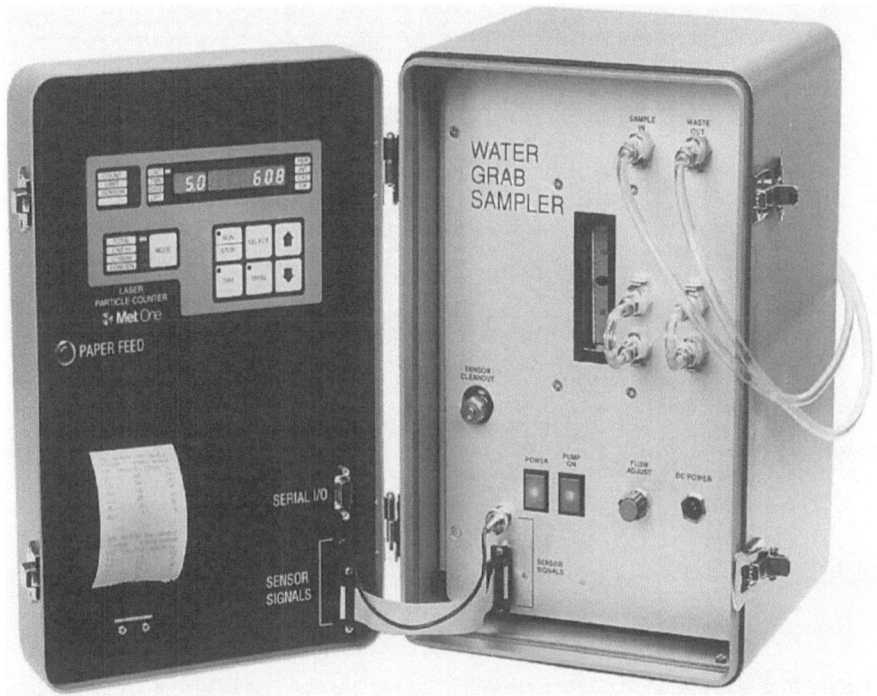

Figure 22.2 Met One WGS 267 grab sampler. (Courtesy of Pacific Scientifc Instruments, Grants Pass, OR.)

The ARTI WQA 2000 comes in a cast aluminum housing with side panels that open to provide access to the sensor sample ports. An LCD graphic display measuring 128×128 pixels is mounted on the top between a membrane keypad and thermal graphics printer. The top is sloped downward to provide a comfortable working angle for the operator. The unit has a retractable handle and shoulder strap, and comes with an optional ruggedized plastic storage case (Figure 22.3).

The IBR VersaCount has a handle attached for portability, but does not have an external case. The front of the unit incorporates a keypad and display, along with a series of bar codes. A bar code reader is attached to the unit. This rather odd feature is included because the unit is a modified version of a Hiac Royco portable air particle counter. Air particle counters are designed to be moved from room to room, since air is difficult to "grab-sample." Bar codes were mounted at various locations where the samples were taken, and provided a quick way of tagging the data. Drinking water plants are not set up the same way, and the considerably heavier water version is not convenient to carry around. It is reported that some of the early customers would stick the bar code wand into the sample, thinking that it was the particle sensor.

The IBR VersaCount is designed such that the sample passes through the enclosure that houses the electronics, power supply, battery, and sample delivery system. A small standalone thermal printer is available as an option.

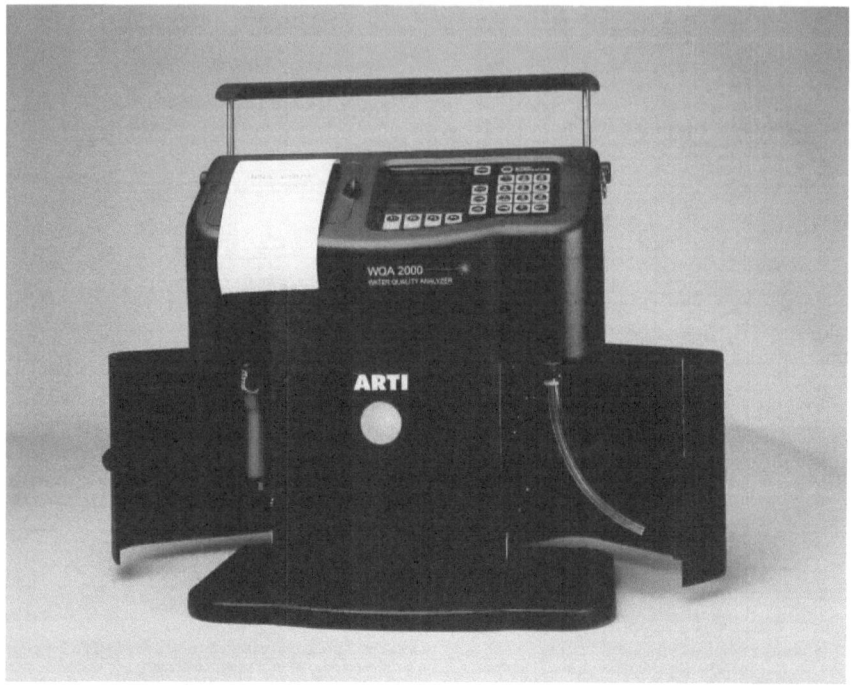

Figure 22.3 ARTI WQA 2000 Portable Water Quality Analyzer. (Courtesy of Art Instruments, Inc., Grants Pass, OR.)

3. Counting Features

The Chemtrac 2400 PS uses the same counter electronics as the standard online unit, but uses different firmware ("Firmware" is operating code that is programmed into the electronics. It is designed to be a more permanent part of the system than "software.") This change allows for up to 16 user-adjustable size channels to be programmed, along with other parameters. Up to 16 tagnames can be programmed by the user as well, so that the data can be stored according to the location where the sample was taken. The keypad is used to cycle through several options for sample time, number of runs, and tagname. Up to 320 samples can be stored in the unit. The liquid crystal display shows the tagname, sample number, and cycles through the particle size data.

The Met One WGS-267 uses a comparator-based counter, which provides six particle size channels preset at the factory. The operator keypad is used to set the sample period and interval, time and date, the number of sample runs, and several counting modes. Total counts, counts per milliliter, or counts per 100 ml can be selected. A special concentration mode provides a quick measure of the particle concentration, so that the need for dilution may be quickly ascertained. An audible alarm will indicate sensor problems, dirty cell, or overconcentration. An ID number can be programmed in to tag the data in memory or on the printout. A log removal function allows instant calculation of log removal between two samples. Up to 250 samples can be stored in a rotating buffer ("rotating" means that the first sample will be overwritten by the 251st sample, etc., so that the most recent samples will be saved in memory if the limit is exceeded).

The ARTI WQA 2000 uses the same counting electronics as the online unit and is available with either of the sensors. The keypad can be used to select from several preset sample and purge volumes. Eight particle size channels are provided, and the user may choose between cumulative or differential counts in particles per milliliter. Samples may be labeled to provide for log removal calculations between the first record of a label and all the other records of the same label. A large storage buffer allows for storing data from up to 500 samples.

The WQA 2000 also accepts inputs for the measurement of other water quality parameters, such as pH, turbidity, conductivity, etc.

The IBR VersaCount provides eight size ranges preset at the factory. Data can be displayed in counts per milliliter or total counts, and log removals can be calculated and displayed. User-selectable alarms provide an audible alarm when count limits are exceeded. The above-mentioned bar code panel is used to input settings. Up to 350 sample runs can be stored in memory for download.

4. Computer Interface

All the grab samplers are equipped with RS-232 serial data interface to allow the stored data to be downloaded to a PC. Both Met One and IBR provide only DOS-based software for this purpose. Chemtrac has a 16-bit Windows program called "Grabbit," which is used to download tagnames and size ranges into the counter electronics, along with a range of sample times and numbers of sample runs that provide the options selectable from the operator keypad. This allows the user to customize selection options. This program also includes reporting software, which is built around the Crystal Reports program used by the online TracWare software.

Met One does have an optional counter electronics and software package designed for those desiring particle distribution data. The counter electronics is a 2048-channel multichannel analyzer, which is mounted in an ISA slot in the PC. A package called WGS is used to collect and view the data. This system is not compatible with Windows software, and must be run in DOS mode. It is designed to take the particle pulses directly from the sensor, bypassing the built-in grab-sampler electronics.

ARTI provides a Windows-based software package for accessing data from the Aquarion.

CHAPTER 23

Particle Counting from a Market Perspective

After evaluating all the available systems and approaches to particle counting, it should be apparent that the "perfect" system does not exist, and never will. There is never any way to be sure that one will make the "right" choice, as experience is the best teacher, but experience is only acquired well after the initial selection.

It is helpful to examine the manufacturer's perspective when evaluating particle counter systems. Certainly there are any number of reasons behind the various ways each company approaches the market, which would be impossible to even attempt to explain. However, human nature and the laws of economics being what they are, a few broad generalizations are possible. It is helpful to understand some of the dynamics involved in the current business climate, especially for those who have spent careers in the public sector, and have not worked in private business.

The municipal market is quite unlike the industrial markets in which most of the particle counting companies have been involved. The long, drawn-out competitive bid process is much slower and more tedious than most industrial contracts. While all human endeavors involve some degree of politics, municipal contracts involve politics with a capital "P," that is, on the governmental level. Decisions are often made with no regard for technical merit by bureaucrats who are only concerned with costs and budgets. New plants can take several years to complete from the time of initial specification, and the manufacturers must invest time and resources in these projects over several years, often to receive no return. Industrial projects are usually accomplished in a much shorter time span, with less political involvement, and the plant personnel are generally more technically adept. The highly competitive nature of private business places a premium on problem solving and process improvement, and once the importance of a particular technology is ascertained, it does not take long to implement.

As municipalities are noncompetitive, there is less urgency to improve. Water quality is quite good in most places, and while lowering the cost of production a few percent would benefit the taxpayer or consumer, it is not a pressing concern. There is no great monetary incentive for investing in a particle counting system.

Market forces are still the driving factor behind the advances in particle counting technology, as in most technological development. Hiac Royco produced a complex and expensive particle counter in the early going, while Met One developed a simpler and lower-cost unit to compete with it in the hydraulics market. This cheaper, simpler approach had limited appeal in the hydraulics market, but was ideally suited for the less-sophisticated and more-price-driven municipal market. As Met One built up a substantial share of the drinking water market against the more expensive Hiac Royco product line, it had little incentive to improve its equipment. This allowed PMS and Chemtrac an opening into the market with systems designed to address some of the shortcomings of the existing equipment. It also allowed Hach to recover from its initial entry with Hiac Royco equipment and try an improved attempt with help from PMS. This has in turn forced Met One to improve its offering, and basically forced Hiac Royco and PMS out of direct involvement in the water treatment market. They have strongholds in other markets, but cannot compete at the price levels that have resulted from this competitive situation.

The importance of competition in this area cannot be overstated. From the early 1990s Met One has been ideally situated in terms of staff and resources to lead the way in developing particle counting technology for the drinking water industry, but has only done so when forced by competitive realities. Chemtrac introduced several innovations, developed by a team of experienced engineers who had been formerly with Met One. Some were hired on a contract basis, and others left after the product was completed, so that none of the development team is still in place. As a result, little change can be expected from Chemtrac in the future.

It would appear that future leadership of the particle counting market will be decided between Met One/Hach and Art Instruments/US Filter. Both have the market presence and technical expertise necessary. Chemtrac is a smaller firm that can comfortably exist on a smaller market share while keeping the bigger firms in line on pricing. IBR will continue to sell to the old Hiac Royco customer base, but is not likely to pose a bigger threat to the others.

But, as the economic skeptics like to say, in the long run, we will all be dead. That and taxes are what we can be assured of. There is no way to know what the future holds for particle counting in drinking water treatment. Enough questions and problems still exist to allow for the possibility of a truly innovative entry that could revolutionize the industry. Unless, of course, firm regulations are established, which will likely end any new innovation in the industry. Competitive markets are what drive innovation, as should be obvious from observing the growth in that most unregulated of markets, the computer industry.

Regulations provide the security of a fixed standard of measure. Once in place, the goal becomes to meet them with the smallest amount of cost, whether in terms of price or effort. People are motivated by incentives, and will follow the path of least resistance to achieve them. When the incentive path is changed from continual improvement to maintenance of a fixed standard, a whole new set of attitudes will be formed. The more positive and dynamic outlook produced by a climate of innovation and improvement becomes a negative and static one of trying not to fail to meet the fixed standard. Failure is punished more than success is rewarded, so the incentive comes to be the avoidance of failure, or even the appearance of failure, more than striving for success. This attitude is prevalent in the water treatment industry, because of the predominance of regulations. It explains much of the hesitancy about selecting a particle counting system, because no one wants to make a "bad" choice. It is safer to defend a choice because the neighboring plants have the same equipment, or the company has been around longer, than to defend it by virtue of its innovative features and performance.

This attitude has become widely prevalent in private business as well, in large part due to the abundance of regulation in all spheres of life. Fear of failure supersedes reward for success, and little incentive is given for those willing to risk failure. Human nature has always been such that outside people, be they competitors or consultants, are always trusted more than one's own employees, no matter how well they have performed in the past. That "prophet is never accepted in his own town" is a well known and established truth. This goes a long way toward explaining why outside competition drives improvement. It is never a safe bet to assume the biggest will always be the best, and it will usually be the newest entry into the market that will push the older established firms to improve. Little consideration is given to customer requests for changes to products until the pinch is felt in the pocketbook.

CHAPTER 24

Preparing Bid Specifications

Most significant equipment purchases in municipal operations require competitive bidding. In such cases, bid specifications must be prepared. It is important that these specifications be written in such a way to ensure that the right equipment is procured at a fair price. This is even more critical if the particle counters are to be integrated into a SCADA system.

A. COMPETITIVE BIDDING

Since particle counting is still relatively new, most new systems are purchased for existing treatment plants. In such cases, the specifications can be written around the desired equipment. When the particle counting equipment is specified for a new plant, or as part of a large upgrade, it is easier for it to slip through the cracks. New plant specifications often place the particle counters in the electrical or instrumentation section. These sections will usually be bid by a subcontractor. In such cases, only an iron-clad specification will guarantee that the desired equipment is procured. Most contractors will know nothing about particle counters, and are only concerned about winning the contract with the lowest bid. They do not have to live with the wrong particle counting system for 10 or 15 years after the fact.

While competitive bidding is theoretically the best way to achieve a fair price and prevent corruption, which is costly to the taxpayer, in practice it can be quite wild and woolly. Subcontractors hold their best pricing until the last possible moment, while the contractors try to make sure that everything is bid to specification. A small mistake can result in the loss of a job worth millions of dollars, or in a costly underbid. A large particle counting system may still only account for 1 or 2% of the total value of the job, so it will command little attention.

There are several ways to minimize the potential for problems when the particle counting system is to be part of a large bid package. The best is to pull it out as a separate bid, to keep it from being lost in the shuffle. If it is to be integrated into a

SCADA system, it should be bid after the initial plant project, once the dust has settled, and the SCADA system has been decided upon. If it cannot be separated out, a prebid qualification is a necessity. This will not only make the contractors job easier, it will prevent a world of problems after the bid.

Anyone who has been involved with bid projects knows that they are always fraught with problems. In the best cases, the consulting engineers have done a reasonably complete job, and the low bidder is a competent contractor who has covered the bases and can make things work when the specs aren't complete or accurate. Change orders can be expensive, and, in the worst cases, lawsuits can result.

It is doubtful that anyone taking the time to read this book will be content to "leave it up to the consultant" or want to deal with things after the bid. While particle counting is coming to be better understood in the industry, the technical aspects of system integration are still not understood well by the particle counter manufacturers, much less the average contractor or systems integrator. We have seen cases where a dozen analog inputs have been specified for the SCADA system to accommodate a dozen serial output particle counters.

B. PREQUALIFICATION AND ALTERNATE BIDS

If the particle counting system is included as part of a larger bid, prequalification is imperative. If SCADA integration is involved, then anything less is bordering on foolishness. In such cases, the best course is to prequalify the SCADA software and system integrators as well. Aside from minimizing the problems outlined above, there are many benefits.

There is no better way to judge the capability and willingness to provide support than before a bid, when the manufacturers are eager to gain the good will of the customer. This applies to the SCADA system providers and systems integrators as well. Let them all work out the problems on their dollar, and not yours. There is a lot less pressure before the bid than after, when the costs are now fixed, and conserving the profit margin becomes the primary concern of all the parties. Once the award is made, unresolved problems can result in costly change orders, as the specifier is now responsible for any oversights.

There may be instances where a particular make of particle counters can be more readily integrated into a specific SCADA package, because of previously written driver interfaces, or other features that may streamline the interface. All other things being equal, this may result in significant savings.

1. Alternate Bids

In most cases, a specification will be written around a particular make and model, since one cannot pick and choose the best features of each. Close evaluation of each of the available systems will usually result in a favorite being selected. Since there are several viable systems now available, there is less reason to try to cold-spec a particular system. This defeats the spirit of competitive bidding, and can result in unnecessary cost.

Depending on the legalities of each situation, alternate bids can be designed to ensure that a competitive situation is maintained. The favored system is specified as the primary bid item, and acceptable alternative systems are listed separately. A price for each system is collected during the bid process. If the price of the favored system is within a few percent of a lower-priced alternate system, it may be selected on the basis of features and performance. However, if the primary system bidder takes advantage of the position with an exorbitant price, the alternate systems provide a viable option.

Alternates may also be used for costly add-ons such as electronic flowmeters. These items may be separated out to keep the system within the budget. It is much better to drop alternate items than to have to rebid the system. Auxiliary items can always be added at a later time, if deemed necessary.

2. Prequalification

When prequalifying systems, it is important to ensure that all the parties involved receive the relevant information. In the case of SCADA integration, the system suppliers will usually be the central figure. They should have familiarity with several types of SCADA systems, and should be able to get the information necessary to interface the particle counters properly. Few consulting firms will have the expertise to design the interfaces properly into the specifications, and in most cases this will be unnecessary. It is important to emphasize that serial interfaces should be used in almost every case, and that the system integrators must be held to this. Since they may not be familiar with the operation of the particle counting system, they may try to promote the 4 to 20 mA approach on the basis of simplicity.

Depending on the size of the project, SCADA vendors may or may not get directly involved. In cases where they are involved directly, the burden should be on them to define an acceptable interface. The particle counter manufacturers will not be able to provide much more than protocol requirements and file-sharing parameters in most cases. They will not want to get further involved, and are not equipped to do so in most cases.

The consultant or plant operator will not have the background to understand all of the technicalities of SCADA interface, but should be prepared to give guidance regarding the type of features and data access that will be required. The simplest way to determine this is to review the features of the standard particle counting software packages, and show them to the SCADA system integrator. While the SCADA software will likely have a different look and feel than any of the standard packages, it should be capable of providing the data in a complete enough manner to allow the particle counting system to be operated effectively. Allow the integrator enough leeway to design the system efficiently, while providing the data in an easily usable format.

The prequalification should be built around acceptable integrators, and they should provide submittals that clearly define their approach within the confines of the project requirements. If multiple SCADA packages or particle counting systems are to be considered, make sure that the integrators provide submittals for all of the options that they propose to bid. Have them include information from the SCADA suppliers and particle counter manufacturers to ensure that they are working with the latest models.

C. AVOIDING PITFALLS

We have already mentioned the imperatives of proper bid preparation with regard to SCADA system integration. But there have been many cases where poorly written specifications for standard turnkey systems have created problems. Badly written specifications leave open the possibility of receiving an undesirable system, or of having to resort to "extra-legal" means to get the desired result. This can take the form of a willful misreading of the intent, or of a biased interpretation of an illogical spec requirement. This will often lead to ill-will or bad feelings on the part of several of the parties involved. Competitive bids naturally result in disappointments, but there is no need to exacerbate them through carelessness. Honest mistakes often occur, but the bid system rarely affords the means for rectifying them. These problems often have longer-term consequences, as a heavy-handed means of getting around a poor specification may result in the offended parties not wanting to be involved in future bids, thus leaving the utility without options for keeping prices in line.

Suppliers usually will take the hint when they are not wanted, and if they perceive that the utility has good reasons for making a choice, will not be offended. If a manufacturer's representative is involved, they will usually want to keep the door open for other products down the line. However, a utility or consulting firm with a history of poor specifications and dealings will get less than optimal response from bidders and manufacturers.

On the other hand, the utilities must be prepared to deal with suppliers who will use less than laudatory tactics in dealing with them. If a particular brand of particle counter is "cold-specced," a competitor may provide a low bid, sometimes omitting key features in the specification, and then try to get approved on the basis of price alone. If they can get to a budget-conscious administrator, they might be able to create problems, either by forcing a rebid to a more open specification, or by getting the administrator to force the operators to accept them. If the specification is not complete and thorough, it is even more susceptible to such problems.

Many systems will start out with only a few particle counters, and then add on a few more each year, or after an expansion. Once the initial system is installed, that manufacturer will be locked in for the future. Without a competitive situation for the next phase, the pricing could rise considerably. Although it is difficult to hold a manufacturer to a price for more than a year, it is certainly possible to require them to bid a maximum percent increase in price per year. Some may be willing to keep the price virtually the same for several years just to get the initial order. Prices have been declining over the years, so a price ceiling is not a great risk for the manufacturer.

Alternate bid items such as flowmeters could be bid in this manner as well. Perhaps they can be budgeted for the following year based on the current bid price. Calibration is another potential "gotcha" that should be quoted in the initial bid for several years. Larger systems may want a service contract, renewable at a fixed percentage increase. Use the leverage provided by the initial bid to secure the best system for the long run, or the initial savings will be quickly lost in the future.

Manufacturer Listing

Analytical Technology (ATI)
680 Hollow Rd., Box 879
Oaks, PA 19456
Phone: 800-959-0299
www.analyticaltechnology.com

ART Instruments, Inc.
1055 Redwood Avenue
Grants Pass, OR 97527
Phone: 541-472-0190
www.artinstruments.com

Chemtrac Systems, Inc.
6991 Peachtree Industrial Blvd.
Norcross, GA 30092
Phone: 770-449-6233
Toll Free: 800-442-8722
www.chemtrac.com

Hach Company
P.O. Box 389
Loveland, CO 80539
Toll Free: 800-227-4224
www.hach.com

Interbasic Resources
P.O. Box 250
11599 Morrissey Road
Grass Lake, MI 49240
Phone: 517-522-8453
www.ibr-usa.com

**Pacific Scientific Instruments
USA** (Met One)
481 California Avenue
Grants Pass, OR 97526
Phone: 541-479-1248
Toll Free: 800-866-7889
(USA/Canada)
www.pacsciinst.com

For up-to-date contact information check the following Web site:
www.ParticleCount.com

Application Papers and Books on Particle Counting

BOOKS

Hargesheimer, Erika E. and Lewis, Carrie M., *A Practical Guide to On-Line Particle Counting*, AWWA/AWWARF, 1995, 129 pp.

Hargesheimer, Erika E., Lewis, Carrie M. and Yentsch, Clarice M., *Evaluation of Particle Counting as a Measure of Treatment Plant Performance*, AWWA/AWWARF, 1992.

Lewis, Carrie M., McTigue, Nancy E., and Hargesheimer, Erika E., *Fundamentals of Drinking Water Particle Counting,* AWWA/AWWARF, 2000, 300 pp.

PAPERS

Andrew, John T., Making meaningful decisions using potentially meaningless numbers: the State of California's experience with particle counting, in *Proceedings 1994 Water Quality Technology Conference*, Part II, AWWA, 1994.

Dunkelberger, G.W. and Musinski, J., Full-scale filtration particle removal evaluation, in *Proceedings 1993 Water Quality Technology Conference*, Part II, AWWA, 1993.

Facey, R.M., Hartery, C. and Gammie, L., Particle count technology for monitoring water treatment performance, pilot study, in *Proceedings of the 47th Annual Conference of the Western Canada Water and Wastewater Association*, Western Canada Water and Wastewater Association, 1995.

Gilbert-Snyder, Paul and Milea, Alexis, California's statewide particle count study, in *Proceedings 1996 Water Quality Technology Conference*, Part II, AWWA, 1996.

Ginn, Thomas M., Jr., Bennett, G. Ricky, and Wheatley, Gregory D., Particle counting in real-world water treatment plant operations, in *Proceedings 1997 Water Quality Technology Conference*, AWWA, 1997.

Goldgrabe-Brewen, Julie C., Count-matched particle counters: experience with quality assurance specifications, in *Proceedings 1996 Water Quality Technology Conference*, Part II, AWWA, 1996.

Goldgrabe, Julie C., Wilkins, Kenneth A., Lai, Hubert, and Marler, Brian, Increasing *Giardia* removal credits through particle removal demonstration studies, in *1994 Annual Conference Proceedings*, *Water Quality*, American Water Works Association, 1994.

Grimm, Michael W., Water treatment plant evaluation techniques: Oregon's experience with particle counting, in *Proceedings 1994 Water Quality Technology Conference*, Part II, AWWA, 1994.

Hunt, D. John, Use of particle counting for water treatment plant optimization, in *1995 Annual Conference Proceedings, Management and Regulations*, American Water Works Association, 1995.

Hunt, D. John, Particle counter count matching, in *Proceedings 1996 Water Quality Technology Conference*, Part II, AWWA, 1996.

Hunt, D. John, Particle counter dilution system, in *1997 Annual Conference Proceedings, Volume A: Management and Regulations*, American Water Works Association, 1997.

Hunt, D. John and Bars, Bill, Particle counter size and count calibration system, in *Proceedings 1997 Water Quality Technology Conference*, AWWA, 1997.

Hunt, D. John and Engelhardt, Terry, Use of particle counting for water treatment plant optimization, in *Proceedings of the 48th Annual Conference of the Western Canada Water and Wastewater Association*, Western Canada Water and Wastewater Association, 1996.

Kelkar, Uday, Opachak, Les, Malloch, Robert, and Jarnis, Robert., Water treatment process optimization using particle measurement techniques, in *1997 Annual Conference Proceedings*, Vol. E: Engineering and Operations, American Water Works Association, 1997.

Koontz, Gene and Shih, Teresa, Filter backwash recycle impacts on the efficiency of particle removal, in *Proceedings 1997 Water Quality Technology Conference*, AWWA, 1997.

Lewis, Carrie M., McTigue, Nancy E., and Hargesheimer, Erika E., Using particle count data in plant operations, in *Proceedings 1996 Water Quality Technology Conference*, Part II, AWWA, 1996.

Lind, Christopher B., A comparison of coagulant programs and impact on particle count reductions in *Proceedings 1996 Water Quality Technology Conference*, Part I, AWWA, 1996.

McTigue, Nancy, LeChevallier, Mark, and Clancy, Jennifer, Findings of the national particle count project, in *1996 Annual Conference Proceedings, Water Quality*, American Water Works Association, 1996.

Myers, Tony, Mejaki, Dale, and Supinski, Anthony, Controlling water plant operations with particle counters, in *Proceedings 1994 Water Quality Technology Conference*, Part II, AWWA, 1994.

Ollier, Laura, Summers, R. Scott, and Bissonette, Eric M., Impact of storage and handling on discrete particle counts, in *1996 Annual Conference Proceedings, Water Quality*, American Water Works Association, 1996.

Routt, Jan C., Arora, Harish, Holbrook, Thomas W., Merrifield, Teresa M., and Peters, David C., A performance comparison of particle counters from different manufacturers: results of a two-year study at West Virginia–American, in *Proceedings 1996 Water Quality Technology Conference*, Part II, AWWA, 1996.

Routt, Jan C., Arora, Harish, Holbrook, Thomas W., Merrifield, Teresa M., and Zielinski, Paul A., Applications and comparison studies of particle counters by West Virginia–American Water Company and the American Water Works System Companies, in *Proceedings 1997 Water Quality Technology Conference*, AWWA, 1997.

Sommer, Holger T. and Hart, James M., The effect of optical material properties on counting and sizing contamination particles in drinking water using light extinction in *Proceedings Water Quality Technology Conference*, Part II, Advances in Water Analysis and Treatment, 1992.

Index